Comedy Matters

Comedy Matters
From Shakespeare to Stoppard

William W. Demastes

COMEDY MATTERS
Copyright © William W. Demastes, 2008.
All rights reserved. No part of this book may be used or reproduced in any manner whatsoever without written permission except in the case of brief quotations embodied in critical articles or reviews.

First published in 2008 by
PALGRAVE MACMILLAN™
175 Fifth Avenue, New York, N.Y. 10010 and
Houndmills, Basingstoke, Hampshire, England RG21 6XS.
Companies and representatives throughout the world.

PALGRAVE MACMILLAN is the global academic imprint of the Palgrave Macmillan division of St. Martin's Press, LLC and of Palgrave Macmillan Ltd. Macmillan® is a registered trademark in the United States, United Kingdom and other countries. Palgrave is a registered trademark in the European Union and other countries.

ISBN-13: 978-0-230-60471-1

Library of Congress Cataloging-in-Publication Data

Demastes, William W.
Comedy matters : from Shakespeare to Stoppard / by William W. Demastes.
    p. cm.
  Includes index.
  ISBN 0-230-60471-4
  1. English drama (Comedy)—History and criticism. 2. Comic, The, in literature. 3. Human ecology in literature. 4. Human ecology—Philosophy. I. Title.

PR631.D46 2008
822'.05209—dc22   2007035814

A catalogue record of the book is available from the British Library.

Design by Scribe Inc.

First edition: May 2008

10 9 8 7 6 5 4 3 2 1

Printed in the United States of America.

Transferred to Digital Printing in 2009

For Mom and Dad and Jean and Erin

A distant end is not an end but a trap.
The end we work for must be closer,
the labourer's wage,
the pleasure in the work done,
the summer lightning of personal happiness.
—Tom Stoppard, *Salvage, The Coast of Utopia* (Part III)

A man hath no better thing under the sun, than to eat,
and to drink, and to be merry.
—Ecclesiastes 8:15

The first thing we do let's kill all the lawyers.
—Shakespeare, *Henry the Sixth*, Part Two (IV.ii.81)

Contents

Introduction: Into the Twenty-first Century   1

1   The Organics of Comedy   11

2   On the Razor's Edge: Between Comedy and Tragedy   29

3   Connecting Mind to Body   53

4   More than Matter Matters   85

5   The Orderly Disorder of Comic Vitality and Its Liberating Potential   103

6   Comedy as Gift   127

7   Comedy Confronts Commodity via the Adaptive Unconscious   153

Conclusion: Things Could Be Better, Things Could Be Worse   179

Notes   185

Bibliography   193

Index   199

INTRODUCTION

Into the Twenty-first Century

"The whole worl's in a state o' chassis," cries Sean O'Casey's Captain Boyle in the 1924 classic *Juno and the Paycock*.[1] Boyle's complaint about the chaotic nature of his native Dublin has a resonance that very well applies to places far removed from his current residence and far distant from his early twentieth-century existence. It seems that humanity just can't get it right, a point that anyone living in the early twenty-first century would find hard to argue against. Despite our unprecedented material productivity, incredible systems of distribution, and nearly instantaneous global communication capacity, humanity remains mired in conflict and fragmentation. The powerful are sold on ideologies bent on silencing and even destroying whatever stands in the way of their ultimate domination, and the remaining mass of humanity remains unable collectively to break out of cycles of disenfranchisement and despair. At best, certain fortunates among the masses find individual opportunities to muster their own personal empowerment over their less ambitious or less fortunate fellow victims. But any universal calls for cooperation, any belief in the value of diversity, or any sense that we're all in it together seem like alien, foolish, and even catastrophic delusions.

The human spirit has been tested throughout history but perhaps at no time greater than during the last hundred years. The persistent rise of almost (though no longer) unimaginably inhuman behavior during this period seems finally to argue that swords rather than ploughshares are the instruments that really separate humanity from the rest of the animal kingdom. If it is true that we are by nature nasty brutish creatures, then the best we can do is to take up arms against our nature and subdue it by force of law, restriction, and prohibition.

However, voices calling for cooperation, diversity, and mutual respect never seem fully silenced even during humanity's countless darkest hours of particular ignominy. There always seems to be some optimistic spring of hope in human nature that never quite dries up, a persistent belief that somehow we *will* finally get it right. Dreamers seem never to cease calling out to us all that humans *should* be nicer to each other, that we *should* be willing to give up individualized gains for the greater good of our fellows. But then we all know that declaring what one *should* do, no matter how sensible, rarely converts gainful human selfishness to doubtful disinterestedness. What we perhaps intuitively know we should do rarely convinces us to do it, especially if there's even a hint of unilateral sacrifice. Humans are rarely convinced of and even more rarely converted to "altruistic" behavior merely through gentle persuasion. Selfishness glorified by the myth of individualism prevails almost as a force of nature in humanity. After all, human culture has done much to naturalize the idea that human aggression is a virtually genetic inescapability. In all such cases, there's this recognition that each of us must look out for number one, or suffer the consequences. And there's nothing worse than losing at this game of mastery and dominion by being a "sucker," by being the gull who lowers his defenses and says something like "let's be friends."

But then again there's yet another way to look at this bleak house of despair, conceding that the current situation may be dire but not necessarily inevitable. If for some unknown reason most of us can theoretically concede (at least in moments of weakness) that we *should* do this or that for the good of our fellows, isn't it possible that there is some inherent absurdity in continuing to behave as the selfish individuals we've become? Could it be that all our problems begin with the mistaken idea that we are discrete, individuated life forms that must struggle against everything other than "me" to survive? It could be that we should recognize our fundamental selves to be environmented parts of some larger whole. And if we succeeded at believing such a point, then maybe we would see that the apparent inevitability of aggression and the normalization of conflict, war, and oppression of others that follows such a concept is nothing more than culturally prescribed delusion.

At its best, revealing this delusion is the comic agenda. At its most ambitious, comedy musters the strength to overcome the delusion. Especially in trying times like ours, comedy really does matter.

Comedy Should Not Be Left Out

This study will subscribe to a perspective on comedy aptly summarized by Joseph W. Meeker in *The Comedy of Survival: Literary Ecology and a Play Ethic (1974)*: "The comic way is not always funny. Humor is sometimes part of the comic experience (as it also is of tragic experience), but humor is not essential to the meaning of comedy. Comedy is more an attitude toward life and the self, and a strategy for dealing with problems and pain."[2] This study will work less to propose a "new" theory of comedy than to highlight how the comic urge in various manifestations persistently struggles to rise to a dominant cultural position that could literally change the world. But how can this attitude that preaches the virtues of the ploughshare while facing the point of a sword ever hope to succeed? When clear reason has failed to convince humanity of the ploughshare ethos time and again, why should we even look at comedy as anything other than a dreamy diversion from real life? How can this "unarmed" approach to life ever succeed at installing forgiveness, mercy, and grace onto the center stage of a culture so sold on dominion, oppression, and retributive justice? These are matters of direst consequence to humanity, but they are too frequently ignored as so much whistling in the wind.

To get a sense of how the comic attitude *may* prevail over its many skeptics, naysayers, and outright opponents, what will be discussed here will be work at the outer reaches of the comic enterprise, the place where comedy stretches sometimes beyond itself and infiltrates into domains typically reserved for other genres. Comedy invades other domains even as other domains frequently overrun comedy, and it is at these ragged intersections that comedy is stretched to its effective limits while revealing perhaps its greatest, most hard-nosed strengths. It is at these intersections that, through comedy, the twenty-first-century mind may best learn the lessons of the comic attitude.

"Learning," however, is not the right term. When comedy works, it actually overrides stern rationality through mechanisms in the body designed to bypass rationalist impositions of control and order. All too often, the "messages" our bodies transmit to our consciousnesses go unheeded, preferring as we do to valorize abstract, mentally generated ideas and to impose their orderly constructs on the disorderly world in which we are forced to live. The result of this strategy is that we often ignore lessons crucial to our survivalist health that are revealed through avenues other than idealized mentality. We tend to sequester those lessons in distant regions of our minds, favoring remedies to our culture's ills generated by our "higher order" faculties unpolluted by local, temporal, and physical circumstances.

As this study proceeds, it will become evident that a cornerstone of the comic enterprise is the body in all its bluntly materialist manifestations, evidenced by humoral physiology, the Rabelaisian/Bakhtinian grotesque and carnival, the marketplace that both sustains and oppresses the body, and the thing called the adaptive unconscious that "speaks" for the body. The body's challenge to rationalist, mental, idealistic orthodoxy is central to what comedy does, challenging abstract and often tragic idealism in favor of materially survivalist alternatives.

Closely related to this point is the fact that comedy has a unique way of looking at nature, marked especially by nature's potential for bounty and excess. This perspective suggests that we turn away from the competitiveness implicit in a zero-sum way of looking at the world. It is not necessarily the case that someone must lose in order for others to gain. Rather, we may be in a position today—perhaps more than ever before—to choose to adopt a non-zero-sum perspective that encourages cooperation for mutual gain. But to adopt this perspective will require the monumental undertaking of convincing selfish humanity that, somehow, its selfishness will be better served not through competition, but through cooperation. The fact that trying to effect this conversion has never succeeded—except perhaps briefly and only within notably small communities—suggests it is a fool's errand. And likely it is. But when we look at comedy and even momentarily suspend selfish suspicions to see nature as comedy proposes, we see a nature at work that creatively utilizes chaotic, nonlinear, and generally disorderly forces of uncertainty and change, forces typically held up as enemies to acculturated, selfish humanity whose systems of order rely so heavily on an economics of controlled commodity exchange. Nature's systems, according to comedy, do not operate in accordance with the orthodox cultural ideal; and that's a good thing. The comic enterprise posits a more benign system of interaction, arguing that unrestricted exchange, antihierarchical order, and nonlinear growth and change are key sources of life as evidenced in nature itself. Rather than selfishly defending oneself against the much-feared ravages of otherness, individual existence may in fact increase its survival odds by working with and actually helping others.

Comedy works beyond, beneath, and around our self-consciously selfish urges for individualized comfort and security. It works hard to give voice to the nonconscious mechanisms of our beings by disrupting the consciously generated mechanisms of hierarchy, dominion, and linearity. In the process, it argues that such disruption will lead not to an even greater likelihood of war and pestilence—as is feared by our encultured self-conscious and selfish selves—but actually to peace and bounty. If comedy ever succeeded at transforming the human world, the aneconomic idea of the world as a non-zero-sum gift would replace the zero-sum economic idea of the world as property, possession, or commodity to be hoarded by the "most fit" (to use a crucial evolutionist phrase). And the idea that we are part of that world rather than some evanescences hovering above it and holding dominion over it—that's the comic vision.

In a world tormented by ongoing threats of war, poverty, disease, and self-destruction, it is reasonable to conclude that comedy has lost its right to be, proof that the comic vision is nothing short of cataclysmic escapism. This perspective is itself a mistake of nearly cataclysmic proportion. Rather than having no place in our culture, I would argue that comedy's place on center stage is more crucial than ever, not to entertainingly divert our attention away from our current lot, but actually to offer a way out of the wilderness of despair we currently find ourselves thrashing around to overcome. Comedy didn't get us into the mess we're in, but it *may* help to get us out.

One important thing that this hard-boiled world of ours has taught us is that comedy, more than ever, needs a spine. It can't simply sing the virtues of its vision but needs to accept the brutality of our culture and prepare means to respond to, rather than capitulate under, brutality's pressures. More than aesthetics depends on this. Comedy, in the end, is as serious as death itself; but it looks to living as capable of creating a barricade against pervasive, pet-rifying fear of this most powerful of oppressors, allowing us to thrive as best we can during our brief time between the womb and the grave.

Kirby Olson, in *Comedy after Postmodernism* (2001), takes a differ-ent approach to matters at hand while reaching many of the same conclu-sions proposed in this study, arguing first the point that "[p]ostmodernism and comedy are aligned in that they function by overturning master nar-ratives and ridding metaphysics of transcendence and closure."[3] Following an agenda advanced by Gilles Deleuze and Jean-François Lyotard, Olson effectively argues that "[c]omedy works by opening rationalism to its sup-posed opposite, irrationalism" (6), creating "a serious immanent alternative to the transcendent philosophies of the sublime and the just" (22). The comic products Olson focuses on deconstruct the monolithic structures of liberal humanism (which prioritizes "greatness" of being) on the one hand, and neo-Marxist justice (which prioritizes enfranchisement of encultured others, including women and minorities) on the other hand. What Olson chooses not to do, however, is to "go from there." He distinctly doesn't see comedy as being able to provide alternatives to the visions it so effectively undermines. But by choosing not to look at the constructive, integrating potential of com-edy, Olson ignores the broader "mission" of comedy. Ignoring such potential, especially in our postmodern (or post-postmodern) world, ignores a large part of what comedy is or does or can be.

What Olson's agenda does do is call attention to a crucial observation pre-sented by Philip Auslander, who—while describing postmodernist political art rather than comedy *per se*—makes the following observation: "Because postmodern political art must position itself *within* postmodern culture, it must use the same representational means as all other cultural expression yet remain permanently suspicious of them. If it is to critique those means by using them, it cannot claim that its use somehow possesses greater truth value than any other use."[4] The problem Olson identifies is that once comedy turns to advocacy of one sort or another, it has compromised its iconoclastic purity. Auslander likely would agree with Olson, but he sees in this "compromise" a threat that *all* active involvement in world affairs must concede. Auslander is suggesting that the art he describes admits a certain unavoidable complicity by being a part of the very cultural practices and behaviors it chooses to critique. It is a strategy evolved from a lesson learned from the radical activism of the 1960s, namely that transgressive "outsider" assaults on the only world we know simply don't work because it is ultimately impossible to step outside or above our own skins and the world at large. We are part of our culture and our world, and by being part we are inextricably complicit in its imperfections. The consequence is that we must undertake more subtle and complex engagements with that world. The result, sometimes, is that the very object of assault *appears* to be valorized.

This cautionary point certainly applies to comedy. Comedy quite frequently is seen as a tool of the status quo, playing with critical fire to a point, but never precisely torching the foundations it critiques. The result, oftentimes, is a sense that comedy, especially those types that include satiric elements advocating conformity and attendant happy endings (a fair number, indeed), actually endorses rather than assaults "the way things are." But we need also to remind ourselves in these trying times that comedy, even satires and those with happy endings, presents a complex transforming dynamic that frequently overrules any trite, affirming-seeming wrap-ups, despite our general tendency to privilege the message of endings. A play ending with dainty dances of fertility may *appear* to reify unsavory patriarchal images of matrimony, while beneath (or before) such an ending, transforming elements may be embedded in the play in ways that redefine the very nature of the matrimony that its ending offers. This sort of subtlety pervades much comedy throughout its long life, and the central problem for comedy frequently seems to involve a need for audiences to look for, to be sensitized to, and to dig for valuable transformations amid a perhaps dubious sense of reification.

In *Bodies that Matter* (1993), Judith Butler argues that culture maintains control and power over alternative possibilities through pervasive *reiteration* of its codes of behavior, perception, and understanding, generating constraints upon what it chooses not to endorse by excluding it from these processes of iteration. She sees Derrida's concept of repetitive "citation" as the process that leads to the cultural normalization of certain concepts, seeing an attendant "sedimentation" that builds up to the point that the given concept becomes "naturalized."[5] From this perspective, it is easy to see the suspicion one would have toward the comic agenda described above, since it would be so simple merely to see the happy endings of comedies as contributing to the cultural sedimentation of status quo concepts and in essence endorsing stultifying conditions of oppression. But if we look differently, perhaps with less suspicion, I suggest that something valuable can be gathered from the inside-out exercise Auslander describes.

Consider that Butler makes the following summary of her position on power politics in general, noting first that "regulatory power [is] reiterated and iterable" and then adding, "To this understanding of power as a constrained and reiterative production it is crucial to add that power also works through foreclosure of effects, the production of an 'outside,' a domain of unlivability and unintelligibility that bounds the domain of intelligible effects" (23). For Butler the world seems inescapably organized by power relations that necessarily—even "naturally"—mandate dominion of one group, entity, or idea over another. For there to be a livable or intelligible condition to exist, its opposite must be identified and presumably abjected, ostracized, or exiled. To change the world, it seems, requires turning the tide on dominant authority—at least provided you're not its beneficiary. Challenging those at the top and taking over is certainly one agenda worth pursuing. The problem, of course, is that such a practice merely replaces one abject other with another.

But moving to another level, one might ask whether or not this presumed system of perpetuating abjection is really a necessary condition for human survival. It may in fact be the case that this underlying system of antagonistic othering, so quickly presumed to be the natural state of things, is itself the result of perhaps the most successful case of culturally naturalized citation/sedimentation in human memory. Rather than believing in the concept of individualized competition, or even competition between and among groups, could it be that collectivized cooperation altogether void of an abjected other might also be a valid alternative? In her study *Witnessing: Beyond Recognition* (2001), Kelly Oliver argues that it *is* a viable alternative, observing "that the dichotomy between subject and the other or subject and object is itself a result of the pathology of oppression."[6] Accepting the subject/object dichotomy as we do, the efforts of psychoanalysis, philosophy, and theology to negotiate truces between the two make perfect sense. In that regard, comedy can sometimes operate to bridge that gap, suggesting legislatable remedies in the bargain. But is it not possible to *re*-vision the world altogether and to discredit the very idea of selfhood and empowerment dependent upon abject othering? In this transformative enterprise, we would look at human development within a cooperative rather than competitive field of operation where others are *extensions* of selves rather than threats to the self.

Comedy sees this transformative enterprise as distinctly possible and, as such, proposes a paradigm in contrast to humanity's very deepest-sedimented obsession with attaining individualized empowerment through subjugation of others. Looking at formerly suspect visions of disorderly nature may suggest how nature itself demonstrates the viability of cooperation over competition. And if what comedy proposes were possible, would we still need to seek patchwork remedies for the symptoms of a pathology that is unnecessarily self-inflicted? Maybe the pathology of oppression is, quite literally, psychosomatic, a result of mental (psycho) self-delusion about autonomy inflicted on and tormenting a body (soma) that constantly wonders why it's being tormented. Convince the mind that things are otherwise than how our culture has sedimented them to appear to be—that we aren't really sick with aggression after all—and our centuries-long search for ways to alleviate the painful symptoms of aggression can be abandoned altogether. Most profoundly, this process entails abandoning foolish idealisms unsupported by material reality. Listening to the body unpolluted by mentally fabricated castles in the air—a genuine inversion of our prior, materially suspicious idealism—is the comic suggestion. This, to my mind, is what comedy attempts.

There's a good case to be made that if we begin the culturally complex transformation of thinking about and seeing the world from a comic perspective, if we allow it to confront and penetrate our currently resistant cultural consciousness, then maybe we can see the beginnings of a new way to build a culture freed from unnecessarily oppressive pathologies. And if we can revision the world in such a manner, we can then also begin a new cultural resedimentation by "citing" terms, conditions, and visions that have for centuries been accumulating in that gently tolerated but hardly seriously beheld comic paradigm. The putative vision of lunatics, lovers, and fools seems at least worth looking into once again.

With the above in mind, this study will view a number of plays from a contemporary perspective, drawing on their relevance to contemporary culture even if they were created in another time. As this book's title asserts, the works of Shakespeare and Stoppard will be central examples throughout, but numerous other playwrights and works will be included as well to highlight why and how comedy deserves a greater place in our contemporary cultural consciousness than it is currently given. What we are looking for is a comic strain that endorses neither a satiric impulse to reinforce the social status quo nor a utopian impulse to dream of reclaiming some ideal state of human nature. Rather, there is a need to locate a hard-nosed comic strain that sees value in negotiating terms and conditions that move beyond the status quo with a view to a more equitable *and* sustainable future.

A Final Note

William E. Gruber is, of course, right when in *Comic Theaters* (1986) he reminds us that the female roles in *Lysistrata* were originally played by male actors.[7] And he is right to point out historical variances determined by Elizabethan predispositions when he summarizes Shylock of *The Merchant of Venice* as originally being an unsympathetic "humorous" figure (4). In the case of both plays, our current sensibilities may in some distant way recognize these historical points, but we nonetheless have made these works "our own." As such, this study will look at "historical" texts but think of the them from a perspective that will—it is hoped—coincide with general presumptions conforming to our current cultural preferences and sensibilities. While there is, of course, no way to confirm singular interpretations of a play as *the* interpretation, there is still some value in striving to conceive of a generally acceptable means of viewing a play, drawing, as this study will do, on efforts to desedimentize culturally embedded misconceptions about the way things are or must be.

The opening section of this study will review the long tradition of comic thought and theory and then will confront works decidedly not comedies even as they utilize decidedly comic elements to confront anti-comic regimentation. Their "failures" as comedies tell us something about the limits of comedy, comedy's fragile nature certainly being implicated as these limits become clear. Succeeding chapters will work to show how comedy strives to reintroduce the body into our decision-making processes and how comedy challenges the stultifying rigidity of human economy in the broadest sense of the term. Finally we will look at comedy's, perhaps unexpected, hard-nosed, pragmatic potential when it comes actually to facing a culture comfortable with comedy's own abjected position in the world of thought, action, and behavior.

The selections that follow are in no way intended to be chronologically ordered, but rather have been selected and ordered in a manner that highlights some crucial aspect of the comic agenda in a manner that follows and supports this study's "argument," rather than in a manner that suggests a historically evolutionary trend (a pattern that Erich Segal's *The Death of Comedy*, for example, excellently though debatably develops[8]). In this regard, the works of Aristophanes and Shakespeare are seen to be as fully relevant to today as Tom Stoppard's latest efforts.

Idiosyncratic as this study may be, the hope here is that at least some of what is proposed resonates with a contemporary readership at least somewhat predisposed to give comedy a chance in these trying times.

CHAPTER 1

The Organics of Comedy

Comedy "wishes to imitate men worse than those of now," and tragedy imitates those "who are better."[1] This is the opening salvo in Western criticism on the topic of comic theory offered by Aristotle in the second chapter of his fragmentary *Poetics*. In chapter 5, he expands: "Comedy . . . is an imitation of the more base, not, however, in respect of every kind of badness, but in respect of that part of the ugly which is ludicrous" (10). Comedy presents men as worse than in actual life, ludicrously ugly, though not precisely morally bad.

Sadly, comic theory has never quite recovered from Aristotle's first fragmentary observations. Though his *Poetics* speaks in some detail to the idea of tragedy, the tantalizingly brief comments on comedy make us long for that section on the comic form presumably forever lost, if in fact it was ever written.[2] But what Aristotle would have said is perhaps less important than the record we have of what he has in fact said, because he and we seem patently in the wrong from the very start—or at least incomplete in our assessments. At first glance, we may be willing simply to concede Aristotle's observation that tragic heroes are better than us average humans, and comic characters are worse. But is that so? Consider the following description by Henry Myers of the tragic hero: "To reach his goal, whatever it may be, he is always willing to sacrifice everything else, including his life. Oedipus will press the search for the unknown murderer, although he is warned of the consequences; Hamlet will prove the King's guilt and attempt to execute perfect justice, whatever the cost may be to his mother, to Laertes, to Ophelia, and to himself; Ibsen's Solness will climb the tower he has built, at the risk of falling into the quarry; Ahab will kill Moby Dick or die in the attempt."[3] The point that becomes evident in this summary is not that these men are "better" than we are. Rather, Nathan Scott observes that the tragic hero, "facing into the utter insecurity of his situation, is led to muster all his resources in one great effort to transcend the fundamental limitations of his creaturehood. It is not, in other words, as Aristotle says that he is better than we are: it is rather that he is, as Henry Myers puts it, more of an extremist than most of us are. . . . [T]hey soon exhaust themselves in the effort to gain release from the restrictions that are a consequence of their finitude."[4] Ultimately, tragedy hinges on a character's unwillingness to accept his place in nature and therefore focuses on his efforts to *transcend* that place. The tragic hero, in short, is unwilling to accept finitude. He (or, less frequently, she) is not merely better than "those of now," as Aristotle says, but is actually better than anyone who ever was or ever will be. Or at least he tries to be. Presumably, according to Aristotle, the striving itself is cause for admiration by those of us who more or less accept our nontranscendent lot. And, presumably, if we don't admire these transcendence seekers, we are probably not too far from being among that group that is "even worse than those of now."

Scott observes that tragedy is predicated on a human urge captured in the writings of Gnostic dissidents of the second century A.D. and is still a major Western undercurrent today, even as it presumably held sway well before Gnostic articulations (for example, in the works of the Greek tragedians of Periclean Athens): "These ancient heresiarchs and *gnostikoi* postulated an absolute seclusion of that which is Radically Significant from all the provisional and proximate meanings of historical experience, and they conceived the world of finite existence to be a delusive and fraudulent imposture. Theirs, in other words, was a god unknowable by nature (*naturaliter ignotus*) and utterly incommensurable with the created order, and man's involvement in time and history was, therefore, felt by them to be a crushing burden and the ultimate disaster from which he was to be rescued" (13). In much the same way that the *gnostikoi* withdrew from the world in order to understand true reality, so have the great tragic heroes of Western tradition rejected truths polluted by time and space in pursuit of transcendent abstractions disassociated from the lowly ways of the flesh.

This urge, however, is not just the urge of the tragic hero on stage. From at least the time of Socrates and Plato, Western civilization at large has cast full doubt on the reliability of nature—and even on the possibility of "knowing" nature—in favor of the ideal forms of which nature is but a pale imitation. Any pursuit that involves natural phenomena as we search for "truth" (except for certain pursuits designed to dominate that nature) has been cast into doubt, the kind of doubt embedded in Aristotle's own valorization of reason over the senses and then by Descartes' privileging of mind over everything other than mind. Knowledge acquired by empirical experience is suspect at best, utterly delusional at worst. If there is an empirical sensory mechanism that is given any credit whatsoever, it is the sense of vision. But vision is given credit and stature primarily because it provides us with the illusion that we are somehow distanced from nature in ways that the "contact" senses of touch and smell are not removed from nature. As a result, vision serves nicely as a metaphor for the operation of rationalist mental "perception" of ideal forms above or beyond material reality: the "mind's eye," looking down on the imperfect physical realm.

Terry Eagleton in *Sweet Violence: The Idea of the Tragic* (2003) moves these observations as they relate to tragedy into the modern age when he observes that "[a]s Hegel puts it in the *Phenomenology of Spirit*, the characters of tragedy are artists, free from individual idiosyncrasies and the accidents of circumstance, giving utterance to their inner essence rather than to the empirical selfhood of everyday life. It is with Hegel above all that tragedy first becomes 'essentialized,' reified to a spiritual absolute which presides impassively over a degraded everyday existence."[5] Eagleton adds, "Mutability, the evanescence of human life, has traditionally been a *topos* of grief, on the curious assumption that what is unchanging or eternal is necessarily to be commended". To escape the limitations of body and pursue pure spirit, mind, and idea—this very much sounds like the standard job description of the typical tragic hero. And this heroic pursuit of the unattainable escape from our physical finitude, since even before Plato and Aristotle, has been the very thing somehow held up as the most noble of human endeavors, if we are to acknowledge the reverence extended to the tragic form when compared to other human expressions of existence, comedy certainly among them.

But perhaps there's another way to look at these "heroes," and a not-so-flattering way at that. Could it be that our headlong pursuit of the abstracted eternal is little more than a self-destructive venture nurtured by our overvaluation of reason? Eagleton points out, "It is one of Nietzsche's many original strokes that he dares to query this doctrine [of the value of the abstracted eternal] and inquire what is so wrong with the fleeting, the transitory, the fugitive" (53). An unyielding critic of our culture's overvaluation of reason, Nietzsche proposed a "gay science" that celebrated present life rather than timeless abstractions. It was a daring proposal, ripe with virtual confirmation of insanity. But it bears attention these days, amid the ruins of rationalist-inspired ideals of "inevitable" human advancement.

Comedy is a suitable place to study the fleeting, transitory, and fugitive dimension of human existence. From this perspective, couldn't the expansive talents of erstwhile tragic heroes have been better utilized in the "comic" capacity of studying the ground beneath our feet rather than the etherized ideal above our heads? Could their talents have been better invested pursuing the pursueable rather than chasing a disembodied ideal that we have no evidence even exists? After all, what is a self without a body? What is an idea like heroism without a correlative object like warrior or attendant verifiable action like confronting a visible foe? And what is the use if nothing is gained (or worse, if all is lost) by taking action against a sea of self-generated, abstracted, and idealized troubles? Pursuing an impossible ideal could be seen to be just so much wasted energy. Maybe discretion *is* the better part of valor, as Shakespeare's Falstaff declares; at very least, it is a concept designed to preserve the flesh rather than to "immortalize" a memory of flesh destroyed in pursuit of the unattainable.

Perhaps the unattainable, after all, does not mark human limitation against which we must struggle so much as it identifies the foolhardy human delusion that contentment with our lot is devoutly to be condemned. Perhaps the unattainable is unattainable because it is something that simply doesn't exist. Perhaps nothing exists whatsoever if it doesn't have a physical manifestation, the "ideal" beyond the caves of our daily existence certainly included. If so, our culture's delusional desire to transcend the physical and achieve the ideal is itself the great human tragedy, so much wasted energy. From this perspective, the ultimate defining ingredient of our being may not be that we "nobly" hunger for an unattainable ideal but that we are part of the very materially real environment that surrounds, engulfs, and supports us. And our aggressive urges manifest in dominating this apparently imperfect sublunary world ultimately amount to a foolish and willful self-destruction of that which guarantees existence. Finally, maybe it's this very pursuit of delusion that has generated this "state o' chassis" that we seem to have created for ourselves.

Comedy's "New" Face

Our efforts to exert dominion over our environment and over "natural" urges have been central cultural obsessions for millennia. The idea that nature must be controlled by culture is a sentiment that inspires a major critical vein in Western culture, even to the point that it infects the ostensible iconoclasm of comic theory, that branch of human expression typically held up as a counterpoint to rational thought, orderly pursuit, and socially "proper" behavior. Consider, for example, that the Aristotelian vision of comedy is fundamentally satiric, targeting nonconforming, "natural" subjects who are "worse than most of us" and shaming those subjects into culturally prescribed societal conformity. Maintaining the cultural standard or norm is what Aristotelian comic theory advocates, polishing the natural diamond in the rough in order for it to shine amid its proper cultural backdrop. Enforcing conformity through the satirical enterprise has long been considered the proper function of good comedy, a sentiment that disqualifies Aristophanic and

Rabelaisian iconoclasm along with numerous other efforts designed to break out of the shackles of restrictive sociality. Comedy, in its satiric manifestation, becomes something of a cultural complement to societal jurisprudence and law enforcement.

Jan Walsh Hokenson in *The Idea of Comedy: History, Theory, Culture* (2006)[6] identifies the Aristotelian/satiric trend in comic theory as a clearly dominant one, seeing the form as a tool of social order dating back to even before Aristotle. However, drawing on the work of Robert W. Torrance[7] and reminding us of various non-Aristotelian surges throughout history, she also identifies a significant nonsatiric, iconoclastic element running just beneath sanctioned social surfaces throughout Western cultural history, a "populist" trend of celebration that challenges cultural orthodoxy to the point of actually satirizing the cultural dominant itself, in stark contrast to Aristotelian formulations. This trend in comic form privileges the individual—the cultural rebel—over the social, a creature who challenges the status quo and, if anything, advocates alterations to the social structure. Never completely silenced throughout Western culture's long history (consider Aristophanes' comic career), but for obvious reasons rarely in cultural favor, this trend has more recently found advocates in critics like Baudelaire and Nietzsche and began to rise to nearly respectable levels in the twentieth century. It's a trend with a doubly monumental task. First, as is the case with all comedy (satiric and otherwise), this populist iconoclasm is faced with the task of securing a voice in a culture determined to (over)privilege the high seriousness of the tragic muse. And then this populist form is faced with the added task of calling into question the very norms and standards upheld by its satiric *and* tragic counterparts, determined as those counterparts are to support and endorse the idealisms of the culture from which they arise.

The goal of this dissenting brand of comedy has been to peel away the dominant presumption that existent social norms were somehow universal features worthy of idealization. Toward the end of the twentieth century, this iconoclastic task took a radical philosophical turn in the form of postmodernism, though it rather surprisingly generally disregarded the legacy of comic populism of which it seems a natural extension. Jean-François Lyotard, among others, challenged the dominant metanarratives of modernism and their predecessors by demonstrating that Western culture's currently held beliefs in a unifying, coherent, or rational order were little more than highly original fictionalizations of reality (whatever those might be). As this massively deconstructive undertaking struck through humanist/rationalist constructions of reality and truth, it left an aftermath of self-doubt regarding exactly where to go following this seemingly consummate triumph of skepticism. Hokenson sees that void being filled once again by the comic enterprise, sobered by postmodernism into thinking and being more precise in its pronouncements by resolutely building on foundations less dependent on rationalist idealisms and more focused on the myriad finitudes that confront us as we strive to reconstruct a sense of self and our surroundings. If, for example, we choose to reconstruct a sense of the social, we will need to lay foundations cemented by something other than ideals of merely rationalist fabrication.

This "transmodern" enterprise of reconstructing a sense of the social returns us to the undercurrented populist iconoclasm that predates postmodernism, but with the added goal of working from the individual toward a sense of the social. After all, merely valorizing the individual is little more than a humanist ideal, every bit as lethal as previous ideals of social normativity. Seeing "mere" aesthetic pleasure as the purpose of comedy (or of any other human undertaking) may appear to be the call of postmodernism, but accepting that turn of events is little more than a fruitless endgame offering no real consolation amid the ruins of postmodern asociality. Jerry Flieger, among many others, observes that human interconnectivity is fundamental to human existence, that even what seem to be merely aesthetic enterprises fall into this dynamic. In *The Purloined Punchline* (1991), he concludes his analysis by observing that perhaps our best decision as individuals is to reject "idle reflection having little bearing on our social practice" and to "take a 'tip' from Beckett's 'waiters,' and adopt a comic position that allows us . . . to see that even the most aesthetic of plays is endlessly performed with and for the Other."[8] Seeing an opportunity to "understand postmodernism as a moment of enhanced social possibility rather than a phenomenon of reaction" (ix), Flieger sees comedy as potentially taking up the task of moving beyond postmodernism's endgame negativity and working toward new formulations of the social in the process.

Susan Purdie's *Comedy: The Mastery of Discourse* (1993), for example, actually returns to sociality by cautioning against a too-strong valorization of the individual over society. Explaining the reasons behind her decision to focus on the linguistic forces of an ostensibly "aesthetic" form like joking, she summarizes, "A study of joking . . . can be seen to highlight the problematic relationship between a necessary and desirable individual empowerment and the destructive exercise of power over others, in a culture where 'power' is primarily understood as identical with aggression."[9] This perspective on what is generally called the superiorist theory of comedy (of which Plato, Hobbes, and Bergson are notable proponents) has clear satirical—and therefore idealistically normative—origins. What is important to Purdie's approach, however, is that she concentrates on the complex roots that ground the sociality underlying the satiric/social premises. Language is here a comic power tool: "'proper' people are those who can produce 'proper' language and, in a reciprocal negotiation, whatever is produced by 'proper' people is taken to define 'proper' language at every level, from the rationality of belief to the orthodoxy of grammar" (18). In other words, whoever succeeds at gaining "the power to define what speaking constitutes 'proper' language" (212) succeeds at defining the cultural norm; and comedy can be instrumental in effecting such changes. Purdie's Lacanian reading of jokes and the jokester reminds us that ideas like "social norms" are, finally, movable targets rather than abstract ideals, and they are unstable especially as one approaches the margins of propriety. It is the joke and the jokester (both in the forms of hero and butt) that test the limits of "proper" sociability. But perhaps most importantly, Purdie reminds us that there is something fundamentally wrong in believing that comedy can or does celebrate individual autonomy: "It seems ironic that comedy—which, if I am right, operates through discursive collusion—should be widely celebrated as a site on which 'the individual' escapes the constraints of 'other men,' and consequently of the inevitably shared systems of language. But I have also argued that the crucial effect of joking is to produce a brief 'mastery' of those systems; and that it therefore, of necessity, occludes its operations" (121). At most, comedy can offer brief individual respite from social bonds. And superheroic declarations of independence are perforce comic delusions as much as they are frequently tragic delusions. What the best of comedy establishes is an ever-shifting terrain of negotiability between the individual and others, predicated ultimately on a recognition of present finitude rather than abstracted ideals.

Present finitude in this case, more than ever, focuses on the embodied existence of the individual, approaching human needs and desires from a perspective that recognizes the body as a (if not *the*) central player in that dynamic known as human selfhood. And everything depends on the continued survival of that body. Joseph Meeker's *The Comedy of Survival: Literary Ecology and a Play Ethic* (1974) and Robert Storey's "Comedy, Its Theorists, and the Evolutionary Perspective" (1996)[10] both return to the Bergsonian species-centered perspective that entwines comedy with vital life forces but with significant modifications. Such vitalist perspectives could, of course, be identified as falling into the category of idealized "master narratives," which was so soundly assaulted by postmodern thought. But rather than conceding that generating master narratives is the result of consummate human delusion regarding the order of the universe, the postmodern critique can instead be taken to inspire a certain restraint in the tricky business of generating cohering narratives. Bergson's use of the natural sciences is a good case in point, and his theory about the comic form is truly valuable today only insofar as his theory is appropriately modified.

Henri Bergson is perhaps best remembered for his two statements drawn from *Laughter* (*Le Rire*, 1900), seeing laughter as "a momentary anesthesia of the heart" generated by "mechanical encrustations upon the living." His vitalist perspective, however, is much more than these taglines imply. Hokenson nicely summarizes the three interactive elements in Bergson's perspective: "Nature is the ground, society is the superstructure, and the individual is the human person who is always evolving in forward motion from egoistic autonomy to group or social life, and beyond" (47). Unsociality, rather than immorality, is the subject of comedy. What the new vitalists have contributed to Bergson is a renewed study of how nature permeates the social through the individual. We know that the individual's biological goal is a rather selfish self-preservationist goal, but science has begun to understand that self-preservation frequently, if not invariably, entails at least occasional moments of mutually beneficial cooperation. At least on occasion, the most effective means of achieving individual self-interest is through cooperative social interaction. Society, then, is not merely designed to perpetuate itself because society is somehow an abstractly "good" thing to perpetuate. Rather it is good only inasmuch as it contributes in ways conformable to the needs and desires of those individuals of which it is constituted.

Bergson's instinct that humanity is naturally a social animal is an accurate one according to current evolutionary theory. His error, however, is that he worked from a species-centered perspective that presumed that humanity functioned for the good of the collective, that we somehow take into account how we would affect species survivability when we made certain decisions throughout our lives. Simple reflection, however, demonstrates this perspective to be patently wrong, given that it inaccurately predicts that human individuals "naturally" consider collective good to be of greater value than individual profit. This sort of consideration frankly exists in no known life forms. While we have effectively generated master narratives that see species survival as a powerful force of nature, there is no evidence that species members think about species survival. It's a simple but significant error in interpreting Darwinism because while it appears that species survival is a motivating factor in nature, what evolution in fact reveals is that individuals are first and foremost selfish survivalists. However (and this is crucial), survival does not always require competitive aggression for success; frequently cooperation succeeds at securing even greater individual survivability than a discretely competitive posture. Society, in short, is "good" not for some reason unto itself, but is good when it improves individual prospects for continued individual existence. The master narrative, in short, that proclaims society as an "ideal" is replaced by local narratives that weigh and determine the relative value of sociability in given, *finite* circumstances. Once this foundational point is established, it becomes less problematic to accept Bergson's vitalism.

Consider Bergson's touchstone observation that laughter is generated when an observer sees humans encrusted with "mechanical inelasticity,"[11] preventing the adaptive qualities of the living being from exerting themselves. The inelasticity Bergson sees in humans is what makes them comic, or laughable, because it prevents the individual from adapting to and fitting into the social roles he is assigned. For Bergson, this comic inelasticity is the result of a lack of individual sociality—rather than ethical/moral shortcomings—deriving from personal "vanity" (172) that leads to the arrogant posture of "seeking to mould things on an idea of one's own, instead of moulding one's ideas on things" (184). Laughter is the communally generated corrective designed to humiliate the asocial or antisocial individual into conformity with societal expectations.

Clearly a satiric perspective on the corrective function of comedy, Bergson's view makes several crucial errors. The first is that somehow humans innately subscribe to the collective/social concept of species survival and that social correctives advocating normative conformity are implicitly good. Second is the assumption that "proper" sociality is synonymous with society itself, that society is somehow properly in line (and up to date) with all that is properly sociable. It is presumed to be a governing institution naturally capable of supporting and enforcing all that is properly sociable in the human species.

What we see here is the general flaw inherent in seeing comic types as either employing satiric correction or endorsing populist/antisocietal individualism. Can anyone today freely accept that society inheres all that is socially good or, conversely, that rugged individualism is a natural human condition? One way of looking at current societal structure supported by satire, rather paradoxically, is that it subscribes to a cult of individualist aggression rather than to the sociality Bergson seems to presume. And often the renegade, antisocial rebel is more an advocate of cooperative sociality than the society he rebels against. Understanding what or who is out of line requires close attention to that thing called the "natural condition"—no easy task.

With the above in mind, what may succeed at redeeming Bergson is a reassessment of the roots of sociability itself. When humans grant themselves the elastic liberties to adapt to their environments (to "mould one's ideas on things" rather than the other way around), oftentimes cooperation is the evolutionary, adaptive result. Bergson presumes that societal conformity grants the human animal its greatest opportunities for elastic adaptation. This is at best a highly suspicious point, worthy of being put to the test by comic protagonists who are elastic enough to test the potential inelastic/encrusting qualities of society on this potentially most social of creatures. This activity is at least as important as satiric testing for encrustations grown on errant, comic individuals.

Yet another point worthy of mention is that the notion of encrustation as laughably comic can also apply to tragic heroes as well. Consider the inflexibility of Hamlet, Macbeth, Othello, and Lear. They have become encrusted with the vain belief that things (the world, other people) must mold around their ideas/ideals every bit as much as the comic hero destined for societal correction. Their vanity is every bit as much the culprit in their downfalls as the comic hero's vanity.

What distinguishes one type of hero from another, according to Bergson, is the degree to which the observer's heart has been "anesthetized." If we are sympathetic, the hero's plight approaches tragedy; if not, he is likely to be laughable and therefore comic. Perhaps this explains why a comic, humorous character like Shylock can approach tragic status given that actors frequently have been able to highlight certain qualities and instill a dose of tragic sympathy into the character. And perhaps a "tragic" hero like Titus Andronicus, whose bloody life on stage tends to numb its audiences, is sometimes regrettably laughable, if not exactly comic. If so, have we not gotten ourselves dangerously close to confusing the two forms merely as a result of *affect*? Isn't there something more to the ideas of comedy and tragedy?

In a perhaps unguarded moment (at least one that seems to sense a problem with the overall scheme), Bergson makes the following observation: "By organizing laughter, comedy accepts social life as a natural environment; it even obeys an impulse of social life. And in this respect it turns its back upon art, which is a breaking away from society and a return to pure nature" (170–71). Here Bergson loosens the grip that his understanding of society has on sociality. Sociality is a force of nature that humanity has absorbed as a survival tool under certain given conditions. It is only loosely manifest in the institution we call society, given that it is a mechanism guaranteed to remain a step behind spontaneous nature *even if* it were a society fully inspired to reflect, support, and endorse the ever-adjusting, fully elastic natural order. From this point of view, *society* is the encrustation that undermines human vitality.

The bottom line is that when comedy best functions, it embraces a non-linearly dynamic interplay between Bergson's three crucial components of nature, society, and the individual. Comedy at its most effective resists the temptation to privilege, say, sociality over individualism because one simply cannot exist (or at least cannot last long) without at least selective interaction with the other. Knowing what one of these components is or means cannot occur without knowing what the others are or mean. And "knowing" either or all requires perpetual renegotiations, unending in their interconnective potential and variety. At its best, comedy engages in this perpetual renegotiation and therefore can only be incomplete if it singularly pursues satiric impositions of conformity *or* individual declarations of autonomy. Forcing the human individual into social conformity without evaluating natural justifications for social formation is every bit as dangerous as idealistically advocating individual freedoms disconnected from the naturally socialized environments in which they best may thrive.

What this means is that humanity *must* listen to its body. *Ideals*, like social order or radical individualism, finally, are unsupportable if not built on realities of physical, bodily needs and if not built on what nature and our natures mandate as optimally crucial for our survival.

Meeker's survivalist, ecological perspective takes us back to that confusing matter of how to distinguish between comedy and tragedy. He observes, "[T]he tragic hero suffers or dies for ideals, the comic hero survives without them," (15) and then adds, "The best thing about comedy is that it is a *way* of perceiving the world and responding to it; a *way* of feeling that is free of sentiment; a *way* of thinking according to wholes made up of clearly recognized parts; and a *way* of acting according to the needs of the context and the tenor of the time. Comedy is a process that proceeds according to its own principles, although it sometimes appears to be unprincipled. In modern terms, comedy is system." (17). Meeker's ecological model is an updated restatement of the Bergsonian model. What it lacks, however, is what Bergson himself lacked: an understanding that this system of cooperation tending toward normative social equilibrium is neither a universally natural given nor can it be generated among humans through rational appeal. Too much of Meeker reads like a species and interspecies contract rather than the tenuous end result of aggressive self-interest securing even greater (though temporary) potential bounty through a momentary though renewable cooperation that can be undermined at any point when individual self-interest is no longer served. In short, a rationally logical plea for cooperation among and within species is an untenable ideal, probably in fact a tragic one for any individual to adopt unilaterally. Espousing the idea that "equilibrium is good" and expecting the declaration to generate consensus conformity simply will not occur. At very least, consensus conformity certainly has not occurred to date.

Robert Storey's more hard-nosed approach recognizes the factual shortcomings of the species-centered argument by drawing on recent empirically supported adjustments to Darwinian theory. He quotes from Stephen Jay Gould: "Natural selection dictates that organisms act in their own self-interest. They know nothing of such abstract concepts as 'the good of the species.' They 'struggle' continuously to increase the representation of their genes at the expense of their fellows. And that, for all its baldness, is all there is to it; we have discovered no higher principle in nature."[12] From Gould's observation, Storey derives two tenets that he sees applicable to comedy: "First, the general mechanisms and manifestations of behavior, including, for *Homo*, not only the emotions but also such reflexes as laughing and crying, have evolved as instruments through which the physical and social world may be mastered. Second, although 'self-interest' may be said to motivate such mastery . . . the 'atomic' self must not be inferred; as Gould points out, the 'struggle' among organisms is the struggle 'to increase the representation of their genes at the expense of their fellows'" (407–8). Indeed, terms like "mastery" and "struggle," implying an "atomic" self-consciousness, really have no place in the discussion given that the idea of "inclusive fitness" engages nothing close to self-aware processes. This idea is modified somewhat with Richard Dawkins' introduction of the *meme* as the mental equivalent of a gene, mental concepts whose "fitness" determines their replication in the minds of fellow intercommunicating organisms.[13] But in all cases of living dynamics, selfishness ultimately rules the day.

The door here is open to see cooperation as a fitness strategy utilized to provide greater assurance of genetic continuation, but it is certainly not an end in itself. And cooperation is equally certainly not a strategy that can be invoked to advocate survival of the species since such a concept simply does not inhere in the natural world.

Storey applies his perspective to the concept of laughter as a survival mechanism, challenging the hydraulic release of tension/energy theory espoused by Freud and others. He then proceeds to discuss various comic types and the strategies with which they strive to master their environments. What, unfortunately, he fails to recognize is that this part of his analysis suffers from similar shortcomings found in Bergson, because his points can apply, with only slight adjustment, to comic *and* tragic heroes. For example, he describes the "rogue" as a mock-aggressive character attractive to the audience inasmuch as his aggression is a game and his actions are play. But does a rogue, even a comic one, recognize that he's merely harmlessly playing a game? Or does he hide behind his ostensible harmlessness in order "to savor the fruits of his audacity" (430) in a manner that describes comic rogues like Falstaff but also mock-aggressive tragic "rogues" like Iago, Richard III, and even Macbeth?

The crucial point here is that much of Storey's biological insights helps to clarify the Bergsonian vitalist perspective and can be used to support an adjusted ecological perspective on comedy. But there's much here that threatens to leak into tragic theory with very little effort. What Storey misses, or at least fails to emphasize, is the degree to which all human actions must be judged in relation to their immediate environment, to the surroundings they attempt to master. Failed full incorporation into the environment against which we sometimes struggle can and often does lead to tragic results. Related behavioral practices under dissimilar environmental conditions can lead to mastery in one instance and to self-destruction in another. Awareness of our embodied selves and how we "fit" into our ever-changing environmental niche is crucial to survival, and survival is the crucial point of comedy.

But curiously, even this is a point that can be gleaned from tragedy. Consider first Meeker's summary of the tragic form: "Our ancestors who invented tragedy some three thousand years ago made a direct assault upon the comic way. What we gained from that was a new sense of human dignity, and a belief that suffering could ennoble us. The tragic view also persuaded us that we could rise above nature and control our own destiny by the power of character and individual. Those powerful messages from the tragic tradition have persuaded us that we need not live by restraints that govern other forms of life. Thus we have become powerful, self-absorbed, and estranged from the Earth" (105). The problem with Meeker's observation—a fairly common one—stems mostly from a belief that tragedy tries to demonstrate the empowering opposite of what comedy does. He misses the point that although tragedy may try to argue that suffering heroically and tragically ennobles, it also suggests that wrong-headed idealism may not be worth pursuing. Tragedy may *attempt* to demonstrate that we can rise above nature and control our destinies, but does tragedy ever succeed at demonstrating our success at such endeavors? We have indeed "become powerful, self-absorbed, and estranged from Earth," but even built into tragedy is a consideration of the cost. To my mind, that's tragedy's ever-present but sometimes unintended point, one that is often missed or willfully overlooked. And at its best, that's comedy's point as well.

Comic Heroes Never Starve,
or the "Manly" Enterprise of Competition

Sadly, our world spends too much of its energy on the tragic longing "what if," listening to the mind's idealized and idealizing disembodied reason rather than to its plaintive earthbound body. And it tends to see tragedy in the way Meeker oversimplistically sees it, as an endeavor to demonstrate humanity's nobility by having humanity resist its apparently lowly place in the animal kingdom.

What is of particular note is precisely how our culture has systematized this antibiological, anti-environmented sentiment by demonizing nature as an other against which humanity must somehow match wits. Particularly insidious is how the sentiment has insinuated itself into our culture's gender politics—supported by myriad historical antecedents—that expressly identifies the feminine/womb/cave as dangerously natural and the masculine/phallic/ideal as devoutly to be valued, even to the point that the latter often must quarantine itself from the temptations of the former. But what if comedy is right in its general valuation of the physical/real over the delusional, abstracted ideal? What if the war of the sexes, for example, is really just one of many descriptive assessments of the mess that the above patriarchally mythologized dichotomy has gotten us into rather than an inevitable, inescapable, and verifiable fact of life? It may be that we don't need to presume that the feminized material world must be placed in a subjective position, othered and thereby controlled as patriarchal urges insist that we pursue some dream of personal, transcendent autonomy.

Kelly Oliver, paraphrasing Nancy Fraser[14] on the matter of cultural injustice stemming from such empirical misperceptions, summarizes two approaches to improving the human condition: "Affirmative remedies for unjust distribution aim at correcting inequitable outcomes without addressing the underlying structure. . . . Transformative remedies aim at correcting inequitable outcomes by addressing the underlying generative framework that produces the inequities in the first place" (50). Affirmative remedies are the remedies that political action seems most capable of addressing, working to minimize through legislation the negative outcomes of our desires to control the physical world so clearly "other" to our idealized being. The brands of comedy that generally fall under the label of satire also rather discursively engage in such surface treatment of cultural aberrations, presenting sorts of dramatized essays on right behavior. Oliver's summary of transformative remedies, however, suggests an avenue running beneath both politics and satiric comedy: "Transformative remedies . . . would redress disrespect by transforming the underlying structure of cultural valuation by destabilizing existing group and individual identities and thereby changing every one's sense of self."[15]

So, if Frazer and Oliver are correct, we really have two options as we confront humanity's current condition. We can either search for surface-affirming remedies that implicitly presume (and therefore extend) the idea that human aggression is inevitably "natural" and therefore requires laws to control that aggression. Or we can strive for the transformative by challenging the very *idea* that is the seed of our aggressive, "unnatural" nature. From there we can begin to find ways to break the cycle, system, or economy of dominion based on these illusions regarding human nature. *Or*, a third option, we can dream the transcendent dream of tragedy.

In the final analysis, the comic perspective argues that the physical world—as imperfectly finite as it is—is our world, and it is better (more realistic, more practical) to try to transform our fish-bowl limitations rather than impossibly to aspire to some transcendent existence beyond the fish bowl, which, *if* we find a way to transcend to that realm, nearly guarantees suffocating doom. From the comic perspective, tragic heroes are ultimately little more than human sacrifices to some misfiring mental mechanism that postulates frustration over the fact that this trivial, imperfect here-and-now existence requires more attention and offers fewer grandiose rewards than we're willing to give and accept.

Is the comic suggestion, this "humble" blue-collar construction-worker agenda, beneath humanity's dignity to pursue? The answer in part is *yes*. We *are* frustrated by such less-than-full-glory agendas because we *do* have problems seeing the virtues of our finite existence. But turning to paralytic statements of despair is not our only option. And neither should we turn exclusively to the noble option of glorifying tragic resistance. In Blank and Jensen's play, *The Exonerated* (2000), one character in particular speaks for many today when he observes, "I know America gets tired of all of these people talking about what they don't have and what's wrong with the country. Folks say, 'Well, what's right with the country?' Well, what the fuck? To make things *better*, we ain't interested in what's *right* with it; we're interested in what's *wrong* with it. You don't say, 'What's *right* with my car?' What's *wrong* with it is what we better deal with."[16]

Admittedly, all is not right with our world—far from it. We live in a post-9/11 world where security threats are a daily concern; diseases ravage whole populations; sexism, racism, and homophobia remain serious problems; and nationalist and ethnic divides result in institutionalized "cleansing." The list of local and global problems is a lengthy one. But recall that the great comic artist Aristophanes lived in a Greece that was devastated by the Peloponnesian War, complete with a devastating plague thrown in for good measure. Shakespeare's England was in a constant state of war and readiness, was regularly visited by plague itself, and famously experienced regular internal strife. And Moliere's world was no less secure and comfortable than either of the above. But although comedy clearly did not transform these worlds in any obvious manner, it may be that these worlds would have been much worse off without comedy around to catalyze changes not directly attributed to or immediately felt by comedy's presence. Despite there not being any clear measures of success, comedy may in fact be more necessary in those troubled kinds of worlds than in those rare ages / places of peace, prosperity, and confidence.

The turmoil of the twentieth and twenty-first centuries may be yet another good example, especially given the unfortunate fact that the pervading tone and tenor of Western art is predominantly dark, brooding, and despairing. J. L. Styan in *The Dark Comedy: The Development of Modern Comic Tragedy* (1968) writes of the late twentieth century: "A renaissance today of the comedy of Congreve and his contemporaries or the confident comedy of Goldsmith, Sheridan, and Oscar Wilde would be an affront to the dignity of the atomic-age audience."[17] If this was true when Styan wrote it in the 1960s (an arguable point), then it should also be true of current times, given even greater cause for concern, insecurity, and anxiety. Erich Segal in his 2001 study, *The Death of Comedy*, echoes the point when he observes that Samuel Beckett's late-twentieth-century theatre is "a deliberate *coup de grâce* to the comic genre,"[18] adding that "the entire Theatre of the Absurd is in a sense a long gloss on Theodor Adorno's famous remark that 'it is barbarous to write a poem after Auschwitz'" (431–32).

But life does go on after (and even *during*) moments of manifest human cruelty and human suffering. Life has a way of moving forward. Even as we are taking up the bodies and clearing a stage (real and metaphorical) cluttered with extinguished life, the living still need to think of a place to sleep and of a next meal. And that is a significant feature of human nature: its ability somehow to survive even in the face of the unimaginably horrific. This point is highlighted by Nathan Scott in a section of his essay "The Bias of Comedy," entitled "Comedy and the 'Whole Truth,'" where he refers to the famous Twelfth Book of Homer's *Odyssey*, following Odysseus' disastrous encounter with Scylla and Charybdis. As expected, the men grieve over the loss of their companions, but they also go ashore to prepare their dinner. At this point, Scott quotes Aldous Huxley's conclusions about this episode:

> [Homer] knew that even the most cruelly bereaved must eat; that hunger is stronger than sorrow and that its satisfaction takes precedence even of tears. . . . He knew that, when the belly is full (and only when the belly is full), men can afford to grieve, and that sorrow after supper is almost a luxury. And finally he knew that, even as hunger takes precedence of grief, so fatigue, supervening, cuts short its career and drowns it in a sleep all the sweeter for bringing forgetfulness of bereavement. In a word, Homer refused to treat the theme tragically. He preferred to tell the Whole Truth.[19]

The point anticipates Herbert Blau's observation that "[e]ven if it's only a radish, comic heroes never starve"[20] because the comic hero's body has decision-making privileges even in the most trying of times. Refusing to idealize existence as the Hegelian view of tragedy summarizes, Homer moves beyond the tragic vision of his earlier *Iliad* to present a more complete picture of human perseverance in the *Odyssey*. Odysseus is portrayed as a flesh-and-blood human being, fully involved with the world around him in ways no tragic hero would ever be portrayed.

The "real" particulars we see of Odysseus' life clutter and cloud the "true" Hegelian abstractions to be gleaned from that life by a tragic poet. Seeing the Whole Truth that comedy provides reveals the full extent of the intricacies of existence that are wholly contingent on a recognition that we are creatures with very particular bodies joined together by a unifying general genetic heritage living in a particular space and time that has common features from other spaces and times as well. For our health, we may very well need comedy even before we tragically sit on the ground and grieve the death of kings and companions.

And for our health, comedy reminds us that we can transform our selves and our environments even as it reveals the limits and consequences of such efforts, arguing that our bodies teach us to think in integrated ways our minds alone have generally abandoned, looking to the abstracted sky above us as the mind's eye does rather than to the solid ground before us as the body's eyes should. And when we relearn thinking itself, we learn that the finite economy of nature is notably different than the zero-sum economies we create in order to regulate its bounty, hoarding it for personal gain. These economies of control prevent an appreciation of nature's expansive, nonzero-sum life-giving, gift-giving potential in favor of some artificial ideal of control and order toward which we too often collectively and foolishly aspire.

Our economies of control, however, extend beyond controlling nature for personal gain. It's been far more embedded into our beings than that. In *The Transmission of Affect* (2004), Teresa Brennan observes, "In theories of psychiatry and psychoanalysis, the healthy person is a self-contained person. This healthy person has established 'boundaries' in early childhood, having successfully negotiated the relationship to the mother."[21] Kelly Oliver observes that these traditional Western theories involving encultured notions of selfhood—or subjectivity—operate within an "economy of recognition" (9) that operates at the expense of objectified "others." As noted earlier, it is a pervading system of oppression wherein dominating subjects subjugate the objects from which subjectivity derives. It is "a deadly antagonistic Hegelian model" (11) that posits a gap between selves as subjects and the objectified world that is utilized to uphold that subjectivity.

Furthermore, this is a dualism fundamental even to contemporary theorists who would seem to be comedy's allies, struggling against pervasive cultural oppression as they do. Theorists from Julia Kristeva to Jacques Lacan to Judith Butler seem most directly committed to generating affirmative remedies to bridge gaps within systems of oppression between the empowered and the disempowered. This is actually something of an unfortunate situation since it generally concedes, as Oliver notes, that we are "forever cut off from others and the world around us do we need to create elaborate schemes for bridging that gap" (12). The perception that we are all a confederacy of disconnected, discrete individuals floating in dissociated isolation from each other is our first problem. It generates a sense that we each have a clear stake in gathering and maintaining personal power over others as we build and defend our subjectivity in this brutal economy of acquisition. However,

Oliver suggests that it is entirely possible that this rather "melancholic" (16) vision of order can be transformationally confronted and overcome by activating counterintuitive leaps in logic predicated, she says, on the notion of love that sees connection to other subjects as natural. It is, she argues, the adversarially based versions of identity that are unnatural, rather than the other way around. Breaking down this economy of domination by recognizing our fundamental human interconnectedness is what she proposes. She speaks, without actually discussing it, to the heart of the comic agenda.

What Oliver and other such advocates of cooperative transformation fundamentally miss, however, is the historically verified fact that rational, discursive, logical calls for such cooperation simply do not work. The flaw here once again lies in a species-centered presumption that the human animal is programmed to accept compromise of our selfish individual urges for the good of the whole. As the biological model reveals, the good of the whole is simply not something built into our biologies. Oliver could possibly suggest that rational debate *may* enact a memic transformation. It could be that by talking up working, living, and behaving for the good of the species could generate a meme that we all may some day be willing to buy into. But even that modest proposal seems a tough one to hope for, given that in order to achieve a cooperative mode of behavior entails working through the fundamental selfishness of our genetic nature.

To get to that point—and to sustain it even for a moment—requires more than good arguments. Convincing the human organism as an embodied being that cooperation benefits rather than hinders survival is what needs to happen. The anti- or sub-rationalism of transformative comedy makes that highly pragmatic conversion its goal.

CHAPTER 2

On the Razor's Edge: Between Comedy and Tragedy

Writing in 1956, Wylie Sypher observes that perspectives on strictly segregated generic distinctions like comedy and tragedy clearly have changed "now that we have lived amid the 'dust and crashes' of the twentieth century and have learned how the direst calamities that befall man seem to prove that human life at its depths is inherently absurd."[1] Sypher's perspective is almost commonplace today, its contemporary pervasiveness having led to the idea that "[t]he comic and tragic views of life no longer exclude each other. Perhaps the most important discovery in modern criticism is the perception that comedy and tragedy are somehow akin, or that comedy can tell us many things about our situation even tragedy cannot" (18).

Unfortunately, those who embrace Sypher's observations have generally succumbed to seeing comedy's "complementary" role still as being rather trite and trivial when compared to tragedy. Comedy variously teaches us to laugh in the face of tragic doom, to cling to isolated moments of happiness, or to tune out the tragic reality of the world altogether. If there is a common formula in any of this, it is that comedy plays outside of the influence of human reason, reveling in the unrealistically fantastic or imaginary and embracing our rationally overpowered lowered faculties. At best, comedy confirms human resilience against insurmountable odds, revealing an admirable survival instinct against all reason. So goes the general argument even today.

But there is much more to Sypher's point that comedy and tragedy are kin or that comedy complements tragedy by offering what tragedy doesn't. The general misconception begins with our culturally held idea that there is a chasm between mind and body and therefore between the "proper" domains of tragedy and comedy. Seeing "serious" tragedy and "playful" comedy as somehow "adding up" to a whole human portrait in fact misses the fully integrated significance of Sypher's point. The interrelatedness of tragedy and comedy goes deeper than superficially "adding up" the two in a yin-yang exercise where the two discretely play off each other by some integrating contrast. The two actually reveal something of the same point.

As suggested earlier, human tragedy more accurately derives not from a failure of reason to prevail in our world but from our adherence to this mistaken mind/body dualistic vision and our current attendant privileging of the reasoning mind over the feeling body. And comedy is more than idle, unminded bodily refuge or unreasoning escapism. Comedy is at its most effective when it argues that the mind/body dualism inherent in our culture and manifest in our discrete (though complementary) tragic/comic visions is itself the delusional source of unnecessary human suffering. It's not a matter of, say, needing to privilege body over mind; it's a matter of seeing that our underlying segregationist urge is actually the source of our problems. Herein lies the root of *both* comic optimism *and* gloaming tragic hope.

Shakespeare's Interrupted Comic/World Vision:
Troilus and Cressida

Of the few playwrights equally adept at comedy and tragedy, William Shakespeare stands head and shoulders above the rest, making him a good starting point when evaluating the interplay of the two genres. Less obvious might be a decision to begin with one of Shakespeare's least comic *and* least tragic plays, but this singular apparent failure-by-design reveals by indirection some of the most significant points of convergence.

Returning to Odysseus, giving him his Roman name (Ulysses), and clothing him in an Elizabethan sensibility, we come to Shakespeare's troubling *Troilus and Cressida*. This play positions us in a world—the seventh year of the ten-year Trojan War—that is clearly at face value unsuited for comedy and therefore arguably not altogether different from how Sypher describes our current age. Because of its unsavory quality, the play ranks as one of Shakespeare's most problematic plays and is consequently often labeled tragicomic. Whatever the label one decides on, however, *Troilus and Cressida* winds up being neither tragic nor comic for several rather telling reasons. Clearly lacking any sense of comic optimism or even tragic hope, the play does uncover roots for a possible fusion of the two by weaving a fabric of inelastic social order against which elastic-hoping human desire must struggle. That humanity fails to gain the upper hand in this play—either through tragedy or comedy—is unfortunate, but that failure doesn't extinguish the possibility that gaining the upper hand is possible.

The characters in the play are engaged in the deadly business of living under extreme circumstances that put their humanity to the test. With few exceptions, the characters are jaded and cynical, negotiating deals and exchanges virtually exclusively for personal gain. Amid these various negotiations, Shakespeare includes perhaps his most precise articulation of the Elizabethan world picture: a grand scheme of an ideal order imposed upon temporal nature, spoken by Ulysses:

> The heavens themselves, the planets, and this center
> Observe degree, priority, and place,
> Insisture, course, proportion, season, form,
> Office, and custom, in all line of order.
> And, therefore is the glorious planet Sol
> In noble eminence enthroned and sphered
> Amidst the other; whose med'cinable eye
> Corrects the influence of evil planets,
> And posts, like the commandment of the king,
> Sans check, to good and bad. (I.ii.85–94)

This complex ideal of the unity of the cosmos is guided by hierarchical dictates in an orderly fashion trickling down from above. While the vision of monarchic hierarchy may not directly describe our own contemporary vision of order, it is based on a logic fashioned by a rationalist's vision of perfect natural and social harmony that parallels a constructed logic longed for by rational beings of virtually every civilized era. That is the ideal, of course, for in the play this ideal vision of cosmic order is almost immediately disrupted. The temporal real quickly fails to conform to the timeless ideal.

Crisis, of course, is a curious constant human condition, which *Troilus and Cressida* reminds us has two potential causes. Prior to Ulysses' lecture, Agamemnon and Nestor have combined to explain why the Greeks' cause before Troy has thus far been fruitless, observing that the seven years' trial before the gates of Troy has been part of a cosmic design instigated by the gods to separate true heroes from pretenders and charlatans. In essence, the current malaise has external causes governed by the gods and therefore beyond mere mortal Greek control. Ulysses' perspective, however, speaks differently. According to him, it is the disarray of the Greeks themselves that has caused the cosmic disarray. The disorderly Greeks are their own worst enemies: "Troy in our weakness stands, not in her strength" (I.iii.137). Conform to the ideal and victory will ensue seems to be the charge. The distinction between the two outlooks is at first glance of paramount significance.

In the case of the former, much of what the Greeks suffer is a result of natural forces (the gods) placing impossible ideals before humanity to try to achieve. A decision must be made either to confront the obstacles and face inevitable failure or to view the obstacles as insurmountable and therefore accept a lesser place in the cosmos as a result. Suffering seems built into nature's fabric given the two unsavory options: challenge and succumb or accept and succumb. In the case of Ulysses' perspective, however, Shakespeare's Greeks create their own obstacles by failing to conform to an apparently *attainable* ideal. Tragedy is not part of the unchanging fabric of the natural world because turmoil is of the Greeks' own making. But neither is comedy inevitable in a world made imbalanced by our unwillingness to conform to the ideal, a point Ulysses himself doesn't quite recognize because he fails to see the flaw in the ideal itself. Given that, as the play (and entire Homeric epic) demonstrates, the system of order Ulysses advocates fails to conform to the elastic necessities of life and living; even proper conformity will not result in a return to or establishment of the order and degree he outlines. So while one vision suggests inevitable resignation or futile defiance and the other implies hopeful possibilities, the remainder of the play demonstrates that this either/or dichotomy is itself conceptually flawed because both visions aspire to idealized visions that inaccurately describe the real world.

Ulysses observes that the willful right hand of this Greek body politic—Achilles—is the cause of Greek discord, and he must be reminded of his place by the Greek leadership. Central to the order that Achilles has disrupted are the warlike virtues of heroism, valor, honor, and chivalry, cornerstones of the grand scheme Ulysses has outlined. Even the apparent chaos of war is part of an order that gives this great struggle meaning. Without these orderly virtues in place and intact, a downward spiral into absurdist meaninglessness seems inevitable. Recalling the Homeric saga, we are reminded that many levels of disruption are at work in the play. If initial disruption of order had not occurred—namely, the rape of Helen—order and degree would not have been disrupted. If the enmity between Agamemnon and Achilles had not torn the fabric of Greek order, the series of disruptive events to follow would not have occurred.

But according to the grand scheme, even these initial disruptions should not be beyond remedy given that the orderly virtues governing human behavior—heroism, valor, and the like—should have minimized the damage of disruption and provided a blueprint for a reestablishment of order. The controlled chaos of a heroic war is in fact part of the natural scheme intended to restore order and degree.

At this point, however, events move beyond the controlling influences of even this blueprint for honorable resolution as Achilles' ignominious slaughter of an unarmed Hector and equally dishonorable public display of his mutilated body rips through the chivalrous safety net and guarantees an even greater level of disruptive brutality that will disorderly run well beyond the events occurring in the play proper, including the doom of Troy and eventual destruction of most of the Greek army. Not even wily, Ulysses appears able ultimately to repair the damage to the orderly ideal. Whether it was the gods or humanity itself that arranged this epic conflict, it is crucial here to recognize that humanity has, first of all, agreed to the conditions that *should* govern its conduct and then, secondly, has failed to live up to those conditions. Why the failure? Was it human failing against impossible odds or simply against attainable ideals imperfectly pursued? The play actually presents a rising suspicion that the ideal order itself—attainable or not—is inherently flawed.

Shakespeare virtually demonizes Achilles in contrast to the noble Hector, vessel of all that is virtuous. But this is not the study in contrast that truly matters in the play, for while Achilles miserably fails to live up to noble ideals, it is Hector's willingness to personify those ideals that becomes the problematic point of the play. Playing either by the gods' rules or by rules established by human contract—it really doesn't matter which is the case—it is playing by these rules that actually dooms humanity to worldly suffering. This is Shakespeare's entirely unorthodox point. He proposes an option that breaks from the traditional, idealized rules of behavior in a revolutionary fashion, one that could, potentially, literally change the "order" of the world itself. And it is noble Hector who suggests a pragmatic, essentially unheroic and even "anti-idealistic" comic option: "Let Helen go." He explains himself,

> Since the first sword was drawn about this question,
> Every [tenth] soul, 'mongst many thousand[s]
> Has been as dear as Helen. I mean, of ours.
> If we have lost so many tenths of ours
> To guard a thing not ours nor worth to us,
> Had it our name, the value of one ten,
> What merit's in that reason which denies
> The yielding of her up? (II.ii.17–25)

Hector decides to face the fact that the price is not worth the prize and never has been. Far from a cowardly conclusion, Hector's decision reflects a comic, pragmatic, and material acceptance of a real/true point and a necessarily simultaneous deflation of the chivalrous and heroic cosmic ideal that so meticulously consumes human beings as it rather inhumanly/inelastically preserves itself. Hector's vital elasticity basically confronts the inelastic code of honor encrusting its adherents. And at this point there is genuine hope for humanity.

In Hector's argument, physical life prevails over ideal honor. Significantly, it comes from a man most draped in honor, willing to give up all in that game simply to return to living. Note that brother Helenus sides with Hector, too, to which a resistant Troilus notes,

> Who marvels . . . when Helenus beholds
> A Grecian and his sword, if he do set
> The very wings of reason to his heels
> And fly like chidden Mercury from Jove,
> Or like a star disorbed? (II.ii.42–46)

Untested, Helenus cannot be the comic voice, for it takes a man of unquestioned capacity to represent such a vulnerable and extreme position. It takes a *hero*, though this brand of heroism has no formal place in the warrior order currently in ascendancy. Having yet to demonstrate a capacity to play the traditional hero and therefore vulnerable to the charge of cowardice, Helenus is not in a position to critique the heroic formula. However, Hector *can* be the comic-hero advocate of a comic resolution, having proven that he can live up to the traditional heroic ideal and therefore illustrating that his *choice* to accept a revolutionarily practical alternative solution is a *real* choice not merely motivated by cowardice. Showing that he can play the game and then choosing the "elastic" option of not playing the game is finally the only way to break out of the self-annihilating cycle into which humanity has been pulled.

Whether this vortex of human suffering has been generated by malicious gods or by arrogant human leadership, the only way to win in this game is to choose not to play by the rules that have entrapped so many in a web of "honorable" retribution and death. That is the comic option Hector proposes, one that requires heroic strength of flexible, vital insight and fortitude greater than that which can be found in any act of traditional honor. Hector's proposal suggests what comedy would comfortably label an act of consummate comic heroism manifest by being willing to pursue profoundly tragic-heroic dishonor. The result is that actual conditions of heroism are inverted in the process by "heroically" challenging the conditions of order and degree themselves.

Unfortunately, Hector is quickly confronted by his hopelessly idealistic brother Troilus, who reignites in him thoughts of family honor:

> Fie, fie, my brother!
> Weigh you the worth and honor of a king
> So great as our dread father in a scale
> Of common ounces? (II.ii.25–28)

Troilus, the conforming idealist, has moved the discussion from the practical matter of Helen's worth (is she "worth" dying for?) to the ideal worth of honor (remember our father), and from there he will later move to the worth of such ideals as love and matrimony. What is ingenious about Troilus' arguments is his conversion of Hector's practical solution to base and ignominious merchant imagery. It is a fairly standard ploy of demeaning an action fraught with greatest sense by equating it with lowly human urges like greed and even cowardice. Though these are by no means Hector's motives, the negative emotional affiliation is such that no one wants to remain an advocate of Hector's suggestion. Even Hector lacks the strength to persevere. In addition to the merchant's scale of common ounces above, Troilus attacks: "We turn not back the silks upon the merchant / When we have soiled them" (II.ii.69–70). Even at the lowly level of mercantile exchange, Troilus argues, Hector has proposed an even more lowly action by offering to violate an already lowly mercantile contract by returning *used* goods (i.e., Helen) to their original owners. Troilus' argument double damns: don't sink to an ethic of lowly market exchange, but if you do, don't go even lower by violating even its lowly rules.

Throughout the play, whenever a practical—and generally "comic"— alternative is put forth, a lowly mercantile equation in contrast to high idealism surfaces to deflate it, obliterating any chance of fairly assessing it as a plan. The value of a pragmatic/comic resolution is obliterated by the very idea that because something is practical it must therefore be dishonorable. Torn between inelastic ideal honor and elastic common sense, Hector ultimately concedes the value of the former, unable to redirect the debate and unwilling to argue in marketplace terms the "value" of a comic resolution. Troilus the idealist wins, drawing Hector (and Troy) unintentionally and naïvely into his idealistic trap in the bargain. Finally, Hector may be strong enough to be heroic in the traditional sense of order and degree, but he lacks the *added* strength to rise to the comically heroic.

Troilus, on the other hand, remains the unwavering idealist. In at least partial defense of his idealism, it is pretty clear that Troilus is mired in an unsavory world motivated by petty greed and self-interest, and he clearly wants more from life than what he sees around him. What is unfortunate, however, is that he has been so immersed in this world of self-interest that he sadly lumps what he sees around him with Hector's genuine life-preserving designs. Unable to distinguish virtue from vice, Troilus adheres to what amounts to a "how-to" code of behavior for all occasions, bringing this idealism to love's arena and becoming inelastically bound to a code of romance that destroys life and living even as it pretends to preserve honor and dignity.

But then, in an additional twist, even this code of love is infiltrated by the merchant ethos Troilus so despises. It turns out that Cressida, the object of his desire, is no virtuous innocent, having developed her own version of merchanting: "Men prize the thing ungained more than it is" (I.ii.301). Working to increase her value by heightening demand, she designs to auction herself off to the highest bidder in this game of love. When Cressida actually becomes a commodity and the Trojans plan to give her to the Greeks, Troilus cannot in turn stoop to engage in negotiations against Cressida's exchange and loss. This man who argued so "nobly" and devastatingly to prevent Hector's idea to return Helen to the Greeks cannot now step into the arena of negotiation because it would involve taking a bargaining position that would be, for him, an ignoble concession to the very lowly self-interest against which he formerly railed. The stigma Troilus, the idealist, has attached to any practical solution has reduced sensible resolutions to trash-heap unacceptabilities. As is the case with the play in general, Troilus is caught in his own noble trap, unable to pursue options that may have resolved his (and Troy's) dilemma because these options are unworthy of the noble posture of chivalric order and degree.

Hector will briefly fight on, gaining one more day of glory; Troilus will lose his Cressida; Cressida will lose her honor; Achilles will lose his honor; the Greeks and Trojans will fight on for another three years to a catastrophic end that need not have happened. Troilus and Hector uphold ideals but lose everything in the venture. Beyond the events of act V of the play, Helen is ironically returned to Menelaus after all, after so much needless bloodshed had Hector's comic propositions been adopted. Add to this the fact that very few triumphant Greeks return home to peace, and the failure of a comic resolution to *transform* the "rules of the games" becomes all the more poignant. The paradoxically noble comic gesture of returning Helen would have prevented a good deal of suffering and destruction, though it would have led to a far less sensational ending to these proceedings.

Failing to achieve a comic resolve, the play fails, too, at exhibiting any real tragic insight. Initially potentially a comic protagonist, Hector promotes the practical/comic resolution of returning Helen to the Greeks, but then he evolves into what should have been a genuine tragic hero, living by and dying because of the ideals he came to personify. Chivalrous even to the point of being at one point warned by Troilus himself against being too honorable, Hector releases an exhausted and vulnerable Achilles from the field of battle to fight another day (when Achilles will be a more worthy opponent) only to have Achilles and his mob of Myrmidon followers slaughter Hector unarmed and defenseless. Hector's chivalrous virtue betrays him, arguing simultaneously for the glory of chivalry and its ultimate idealistic impracticality in this gory world that finally does not support the ideals Hector so ably epitomizes.

But unlike actual tragedies, Hector's too-noble tragic conflict and ensuing tragic demise do not result in any return to order, certainly not to the sort of ideal order articulated so well and so fully by Ulysses back in act 1, scene 3. Rather, the play ends bitterly and without resolution of any sort. The twisted and thwarted Pandarus speaks the final words, referring to prostitution ("the hold-door trade") and venereal disease ("some gallèd goose of Windsor") as he exits:

> Some two months hence my will shall here be made.
> It should be now, but that my fear is this,
> Some gallèd goose of Windsor would hiss.
> Till then I'll sweat and seek about for eases,
> And at that time bequeath you my diseases. (V.x.52–57)

This world is left hanging at play's end in the clutches of uncertainty and cynicism. Love actually does sink to the level of the marketplace along with attendant disabling lust and disease. Chivalry's reward is death, constancy is made a fool, and comedy is humiliated into silence.

The unsettled nature of the play's conclusion stirs questions regarding genre. The problem is not helped, either, if one opens up the categories, Polonius-like, and considers the category "tragicomedy" or even "dark comedy" (Styan's term). For one, both tragedy and comedy, classically speaking, pursue a common goal: moving us into and then pulling us away from disorder by returning to an old order or moving us forward to a new one by play's end. *Troilus and Cressida* moves us into disorder, but the enterprise ends at that point. As such, the issue should not so much whether it is tragedy or comedy by why it is neither. Rejecting comic resolution as Troilus has done steers all toward tragedy. But in this play, even tragedy fails properly to arrive. Tragedy begins to unfold in the form of Hector's rise and fall, but nothing comes of it. Love is left in the mire of inconstancy and cuckoldry, with no attempt to resurrect it. We neither find transcendent "tragic" value nor a transformative "comic" potential in the play. Hector's comic suggestion and tragic doom, Ulysses' manipulations of various contingencies, Troilus' rejection of comedy altogether—the play is a tale of multiple failures leaving little hope for resolution beyond the mere "end" of catastrophe.

Remembering that *Troilus and Cressida* is formally labeled, not surprisingly, a "history" reminds us of Langer's point about history, that it "meets good and evil fortune on countless occasions but never concludes its quest."[2] But simply conceding that *Troilus and Cressida* fits into Langer's description of a history play doesn't really end the matter, because in the case of this play at least, it is worthwhile to uncover exactly why resolution—tragic *or* comic—doesn't occur. The problem centers around how the elastic practicality of comic invention is cast and then cast aside. The nobility of the heroic is presented well in the person of Hector. And Shakespeare does suggest a comic resolution that virtually all students of the Trojan Wars have likely pondered: why not just return Helen to the Greeks? By putting such a proposal in Hector's mouth, Shakespeare has given his play's greatest character the opportunity to *elevate* this traditional (perhaps even tragic) hero to the status of comic hero.

What happens to this option reveals Shakespeare's genius in this troubling play. The unrepentant idealist Troilus assigns all "practical" action the lowly status of merchant-class behavior, which is by all accounts simply motivated by greed and self-interest. In this warrior-class world, assigning anything to the level of the merchant class quickly reduces it to poisoned fruit. And by extension, anyone associated with a practical idea designed to arrive at a negotiated settlement—Hector certainly included—is a person motivated by little more than cowardly self-preservation and self-profit. So when compared to the ideals of the tragic hero, comedy comes out as an utterly unsavory alternative, something completely lacking in noble gesture. From this perspective, the comic resolution is clearly beneath "noble" human nature to pursue.

But it need not necessarily be the case that comedy is a selfish, lowly enterprise looking, literally, to capitalize on exchange for personal profit. It is not necessarily the case that comedy's investment in such physical qualities as self-preservation reduces comedy to the lowly level of pandering for self-gratification and hollow, selfish creature comfort. There is the possibility of seeing this investment as something of a noble position itself, in some ways even more noble than the boiler-plate path to nobility outlined by some stultifying code of chivalry. Preventing the utter destruction of whole cities and cultures, after all, seems a worthy goal, especially if the plan includes what many agree would have been righting an earlier wrong.

Here we reach yet another critical point. Consider the high likelihood that trying to return Helen in this play probably wouldn't have succeeded anyway, given that one could argue that the Greeks' true mission was one of conquest for personal gain (the riches of the East) that was merely *cloaked* under the idealistic and "noble" guise of defending the honor of cuckolded Menelaus. It is entirely possible that the high ideals that seemed to have inspired this epic struggle were mere smoke screens for less honorable motives like greed. Hector's offer to return Helen—if actually extended to the Greeks—could very likely have merely revealed the Greeks' "honorable" intentions for what they actually were. So close to capturing the golden city on the hill on so many occasions throughout their campaign, the Greeks very likely would have found a way "honorably" to reject the offered return. Instead of this being a war intended to recover Greek honor and order, it is more than possible to see the entire ten-year "heroic" enterprise to be inspired by lust and greed, a matter of lowly avarice only *veiled* by honorable ideals.

So it is actually possible to take Troilus' argument against comic self-interest and use it to deflate heroic nobility in much the same way. Heroic nobility can be shown (at least on occasion) to be nothing more than a well-made cover for less noble interests in virtually the same way that comic solutions can be shown to be motivated by the same. If we even for a moment consider accepting Troilus' view that greed, avarice, and cowardice are the motives behind comic resolution (at least sometimes), an honest assessment of tragic acts of nobly heroic behavior may arrive at the same underlying motives.

But comedy is surely more than the lowly pursuit of creaturely comfort and escape from danger, though Troilus' reductive charge is one that comedy often finds difficult to separate itself from. Consider the fact that the failed suggestion to return Helen is not necessarily lowly or cowardly, being instead a frank concession that a wrong has occurred and should be rectified. The comic "nobility" in the suggestion lies in accepting responsibility for an earlier act of injustice, demonstrating at very least that comedy is something more than the unacceptably negative and polar opposite of the incontrovertibly positive tragic/heroic ethos. And aren't human relations better served when a wrongdoer freely accepts responsibility and withdraws from the lists on his own? Here we have the clear existence of a flexible and even noble sense of justice in the fundamentally comic suggestion belying the general claim that comic resolve involves an abandonment of honor, barely veiling its fundamental covetousness and cowardice.

So *Troilus and Cressida* presents a nearly comic hero and a partly successful tragic hero in Hector. But in the end, comedy is humiliated by the company Troilus claims it keeps and by Hector's unwillingness or inability to defend it against such defamation. And any tragic resolution to be won by Hector's death fails to arrive. Conceding a need to break from this death-dealing system of order and degree doesn't happen. What we see here is that the trademark traditional resolution of tragedy is actually grounded in a comic sense of ultimate (though delinquent) compromise and adjustment, a "purchase" of order and continuation of life, albeit the tragic purchase is far more costly than need be, calling for the blood and suffering of innocence amid seemingly needless intransigence. But even among the ruins, finding peace in a system-breaking resolution of this sort smells too much like the comic capitulation Troilus impugns, and so Hector's death is just so much waste, with nothing learned and nothing gained.

In *Troilus and Cressida*, tragedy is unwilling to concede the inevitable triumph of comic design and therefore fails to be either comic *or* tragic in its end. And as a result, neither tragic transcendence to order or degree nor comic transformation of that order and degree occurs.

The Comic-Tragic Folly of Hamlet

In *Troilus and Cressida*, comedy and tragedy grimly stand toe to toe, slugging it out on center stage until neither is able to answer the final curtain. No winner can be declared, but there is a loser, and that loser is humanity. Much of the misery in the play stems from the false dichotomy that surfaces when two apparently contradictory visions on how to deal with the world run up against one another. Erich Segal succinctly summarizes this point when he observes that the "tragic hero dies for what is nobler in the mind, the comic hero *lives* for what is humbler in the flesh."³ Implicit in his statement is the unfortunately mistaken point that mind and body operate in a sort of disjointed manner, never quite on the same page and in fact not very often operating in the same plot or even on behalf of the same organism. It is a point that has tormented humanity for centuries, triggering momentous self-inflicted catastrophes against which even the archetypal Trojan War dims in comparison. And it is a real-world point that finds its theatred double on the stage in the form of the struggle between tragic and comic visions, bloody when tragedy prevails, something altogether different when comedy succeeds, and catastrophic when neither arrives.

In *Hamlet*, a similar struggle erupts between comedy and tragedy, but the results are different this time because here Shakespeare manages to challenge the concept that such a mind/body chasm necessarily exists. Here we see a distinct possibility that the two realms could come together and move humanity to a level of circumspection that reevaluates presumptions inherent in the terms comedy and tragedy themselves. Shakespeare moves onto this plane of consciousness in *Hamlet* and takes us surprisingly close to a comic resolution among generally tragic circumstances. Tragic resolution does prevail, and though there are clear losers in this work as there were in *Troilus and Cressida*, what evolves is a hope in tragedy that comes surprisingly close to echoing the optimism of comedy. Because tragedy does win out, humanity profits in ways that *Troilus and Cressida* never fully allows, though the winner (humanity) pays a dear—and sadly exorbitant—price for the victory. If only comic resolution had won the day.

Hamlet has all the idealistic trappings of Troilus but is far more intriguing, given that his greater self-awareness leads to what can arguably be described as an actual comic transformation. By all accounts, he has even greater potential than Hector to turn the tide toward a more-or-less bloodless/comic resolution. But he does start out as someone more like Troilus than like Hector. Hamlet posits his initial stance in his first entry on stage when he declares, "I am too much in the sun" (I.ii.67), a compact declaration that is revealing on multiple levels. He regrets being an embodied creature dependent on the physical, external energy of the sun to sustain itself, desiring instead to be something like a self-contained autonomy. He despises the fact that he can be called "son," hating that he has a body spawned by the physical union of a mortal father and mother. As a dependent creature on several levels, he is painfully aware that he is susceptible to the physical transfer of allegiances manifest in becoming a stepson. And he hates that events have transpired to place him roundly in that sun's/son's spotlight, confirming his physical reality by way of the now unavoidable gaze of all around him. He even shortly proclaims a desire that his "too, too solid flesh would melt, / Thaw, and resolve itself into a dew" (I.ii.129–30). Trapped by the putrefying inescapability of his flesh, Hamlet is clearly a man who prefers the ideal life of the mind found in Wittenberg rather than the carnal intrigues of Elsinore or even the cosmopolitan pleasures of Paris. And all of this occurs before Hamlet is even apprised of his late father's dastardly murder. So it's no surprise that he curses his "fate," the depressing and potentially tragic point that *he's* been born to set this corruptible physical world right. Disinclined even to be in this world in the best of circumstances, he is doubly burdened by having to correct the matter of his father's untimely death before he can even begin to contemplate the possibility of addressing the timeless matter of the world's inherently corrupt condition.

Charged by a specter to avenge his father's "foul and most unnatural murder" (I.iv.26), Hamlet famously hesitates, clearly unwilling to sully his hands with the world around him, even to the point that he disdains involvement with events and agents clearly not involved in the murder. Hamlet shows nothing but contempt for a dundering Polonius who is so proud of his physically regenerative good fortune—in the forms of Laertes and Ophelia—and of his administrative place in the world in general. He has contempt for Claudius even before he has suspicion that this new king/father is a regicidal/fratricidal tyrant, a man wickedly hungry for a place in the maggot-breeding sun of the physical world. Lesser luminaries like Rosencrantz and Guildenstern likewise hold no favor in Hamlet's mind's eye. And Gertrude and Ophelia are tantalizingly dangerous sirens of physical attraction for Hamlet as he struggles to create and maintain objective distance from this world so bent on ensnaring him and dragging him from his idealistic perch. The very quality of their attraction necessarily repels him and compels him to use force to reject the emotions that try an embrace designed to enfold him into creation and that threaten his world-distancing autonomy.

Ironically, it is Hamlet's continued resistance to the physical attraction of the other—in the form of other humans and in the general form of the "othered" world at large—that has soiled this once spotless personification of the ideal. He is in an impossible situation wherein he can either accept the world's embrace and sink into its folding arms, or he can resist and fall prey to an aggressive posture of self-distancing that will inevitably stir passions of contempt, loathing, and even self-hatred. Maintaining distance from this world, it appears, is impossible. The world has hopelessly trapped Hamlet.

All complications, snares, and traps notwithstanding, Hamlet chooses to move forward. Despite his first-act reluctance—"O cursed spite / That ever I was born to set it right" (I.v.188–89)—Hamlet accepts that he is born to set things right and dives into an ethical, epistemological, and ontological cesspool, seemingly fully equipped to reverse the pattern of decay and rot plaguing Denmark. Bolstered by a mental library of modern ideas, Hamlet allows his mind the free play to step out from the stale shadows of the Middle Ages and bask in the infinitely ideal (though not quite accurate). Copernican sun of unlimited human reason. He is at this point a man entering the lists representing humanity with a fully steeled mentality ready to withstand all the slings and arrows that may come his way, standing upright and ready to dodge or absorb those blows full on.

The primary field of conflict in these matters, of course, involves the instability and perceptual duplicity of the physical world, its ability to seem other than it is, a problem nonexistent in the ideal world of mind. Hamlet—or more precisely, Shakespeare—was certainly aware of how deviously the physical world can be manipulated against the idealized true, given that such actions were systematized and presented to maximum positive effect by Renaissance contemporaries like Machiavelli (*The Prince*) and Castiglione (*The Book of the Courtier*), among numerous others. The quality of *seeming* to be true or real in the physical world could never inhere in the certainty of the unshrouded *being* qualities that exist in the unencumbered world of the ideal. Physically seeming virtuous, for example, and actually being virtuous could never fully be trusted to coincide in the world of the physical.

But at least early on, Hamlet is not put off by such dubious qualities in the physical world. Trusting his mental superiority and the certainty of its/ his moral compass, he confidently dons unadorned black garb and declares an independence from and superiority over the illusory world of materialist seeming, scoffing at the very idea of seeming that is introduced in the play by a troubled and confused Gertrude ("Why seems it so particular with thee?" [I.ii.75]). If any idealist can remain uncorrupted by the physical world, it should be Hamlet, possessed as he is of virtually every ideal virtue available to humankind. Aware of dualist traps set out against arrogant and unwary humanity, Hamlet catalogues for us a list of human affectations:

> Nay, I know not "seems."
> 'Tis not alone my inky cloak, good mother,

> Nor customary suits of solemn black,
> Nor windy suspiration of forced breath,
> No, nor the fruitful river in the ye,
> Nor the dejected havior of the visage,
> Together with all forms, moods, shapes of grief,
> That can denote me truly. (I.ii.75–83)

He then notes that

> These indeed seem,
> For they are actions that a man might play,
> But I have that within which passeth show;
> These but the trappings and the suits of woe. (I.ii.83–86)

Neither a Machiavellian prince nor a Castiglionian courtier, Hamlet asserts that he is what he seems to be, no more, no less, because striking through the trappings and suits of seeming is not a serious problem for a man of his superior vision. The line is clear and unambiguous in Hamlet's mind: accepting the point that dualism (the physical and the ideal are separate realities) is incontrovertible, he concludes that the physical/material world is little more than a poor shadow of the Platonically ideal real (or the Christian truth). And seeming and being are clearly distinguished by a shrewdly trained mind such as Hamlet possesses, unpolluted by "things rank and gross in nature" (I.ii.136). Hamlet is a man who knows the difference, and if necessary (though quite reluctantly), he can operate in both worlds, calling his ideal mind to grapple with and ultimately unmask the rank material world that surrounds it. He prefers the ideal but feels more than adequately equipped to wade through the too, too solid world of the physical if need be. This, at least, is where Hamlet stands at play's beginning.

As the play unfolds, however, it becomes increasingly clear that it is nearly impossible for Hamlet to distinguish between seeming and being, despite his claims to know the difference. As Hamlet moves through Elsinore, seemingly in control of his outward behavior, questions begin to arise that even he appears incapable of answering with any certainty. Put another way, Hamlet's actions and the play in general at first support the standard concerns that the material/real does not coincide with its ideal/true correlatives. The physical world is not to be trusted to reveal anything of certainty about the ideal/true world. But in that real world of confusing seeming/being, can Hamlet really tell the difference as he claims, positioned as he believes himself to be from an objectifying perch above the material goings-on that he "playfully" steps into and then withdraws from? Distinguishing between claims and actions, can Hamlet (or even the audience, for that matter) draw sufficient distinctions between seeming and being to accurately answer the following (just to list a few disturbing questions): Does Hamlet love or hate Ophelia? Does he or doesn't he want to avenge his father's murder? Does he love Gertrude? Does he have any true friends in Elsinore, or elsewhere? Is Hamlet a truly disinterested man, or are his actions motivated by elements of self-interest? The truths, of course, are cloaked in ambivalence and contradictory signs, but they're by no means utterly unfathomable.

Shakespeare presents a Hamlet who eventually sees exactly how his mind has run before his body thanks to unsupportable presumptions. Deceived by sophisticated dualistic abstractions claiming the primacy of the mind over the body and the ideal over the physical, Hamlet (catching up to Shakespeare himself) evolves, not from a high-minded idealist like Troilus into a cynical realist like Pandarus or Cressida, but actually into a person who dissolves the realist/idealist distinction altogether and becomes what could be called a monist in his way of thinking about the world. The idealism gleaned during his otherworldly days surrounded by books and ideas in Wittenberg is challenged by the high physical theatricality of Elsinore court life. And this clash of worlds leads to a transformation in thinking that fundamentally alters Hamlet's vision of reality. He comes to realize that the real/physical and the ideal/mental are almost inexplicably but nonetheless unavoidably inseparable. What this means is that Hamlet ultimately comes to understand that seeming does not occupy one realm and being another; rather, within this one world, the matter of sorting out seeming and being—though of considerable *pragmatic* concern—is ultimately a matter of engagement with and disciplined perception of the world that surrounds us. Life and living become a sort of detective's job of deciphering a wide range of events and evidence within a single, fully physical sphere of action. It is no longer a matter of seeing existence as somehow coexisting in two separate spheres as Hamlet's earlier dualist musings presumed.

Hamlet, of course, is not alone in struggling with such matters. From at least as far back as Plato/Socrates, the physical realm revealed through the senses was demonized in favor of the ideal forms of which the physical was at best a poor shadow. At heart is the struggle to strike through what Jean Baudrillard describes as the simulacral realm of appearances, an agenda Western culture struggles with at virtually every turn even today. But implicitly following a tradition inaugurated by Émile Durkheim (in *The Elementary Forms of Religious Life [1912]*)[4], Baudrillard suggests that we not abandon the simulacra in favor of ideal forms; rather, we should search for distinctions between what he calls "authentic" and "inauthentic" simulacra[5] as we search for a way to reclaim a certain "enchanted simulation"[6] that could reveal and reclaim a "charmed universe"[7] that seems otherwise available only in the fanciful and elusive world of ideal forms. Exactly how such an enterprise should be pursued is itself something of a mystery, perhaps requiring faith or some other non- or pre-rational mechanism.

But Hamlet's transformation doesn't require purely mystical explanation. As the play unfolds, Hamlet's agenda increasingly goes off track primarily because he has chosen to limit his resources and uses reason alone as he seeks truth. The hints of another option are evident in Hamlet's frequent and ironically emotional contempt of himself for having become emotionally engaged with the world at various stages throughout the play. Joseph Meeker, for example, points out that Hamlet's famous indecision stems in large part from beginning to sense that he lives in "a biological environment that he shares on more or less equal terms with the other animals."[8] From this biological position, Meeker argues that Hamlet taps into an ennatured sensibility that generally resists death-dealing encounters in favor of more ritualized behaviors. In other words, Hamlet resists avenging his father's murder most foul as a result of a sort of natural resistance to mortal combat inhering in nature in general. Despite "knowing" that honor requires vengeance, Hamlet "feels" the natural calling that repudiates such action.

Consider that while aggression in the natural world clearly exists, it is generally redirected through "special inhibiting mechanisms" (Meeker 40) that prevent wasteful and destructive mortal combat. But taking into account that humans needed to *develop* tools to kill in ways the rest of the animal world did not, Meeker notably adds, "The human capacity to kill is mostly a cultural invention, not a natural gift, so we lack the instinctual controls over our weapons that other animals have" (41). What this suggests is that while defending one's territory or honor may have natural roots, the urge to defend unto death is a nearly uniquely human, culturally generated phenomenon untempered by the naturally "con-speciated" development of aggressive redirection that inheres in most other species. So, rather ironically, evolutionary biology teaches us that the urge to kill among humans is not a primitive, natural, or "uncivilized" urge so much as it actually is an encultured—and inaccurate—extrapolation of how humans would behave in a "state of nature."

So Hamlet's inaction seems at first glance to be nothing less than cowardice. Then his *seeming* natural "instinct" to defer revenge appears to be the result of Hamlet tuning into his better "nature." But following the complex intertwining of natural and cultural influences in this matter, we see that Hamlet is actually following a path that must somehow recognize that the human cultural ideal of mortal revenge comes with a well-grounded counterurge to avoid mortal combat that is hardwired into other "natural" killers in the animal world. From Hamlet's idealist perspective, however, such urges to redirect his aggression as nature would recommend—over what his idealized, encultured code of honor mandates—are actually evidence that he's fallen from the high road of reason, the only path to truth as he sees things.

Confused by the complexity of the above influences and personal reactions, Hamlet understandably falls victim to self-contempt. Hamlet's self-contempt, however, will shortly change. It is a change that moves Hamlet away from the tragic and into the comic realm by way of an attitudinal adjustment that curiously anticipates advances in the cognitive sciences that have arisen in recent decades. Antonio Damasio, neuroscientist and author of *Descartes' Error* (1994), observes the evolutionary fact that "[i]f there had been no body, there would have been no brain"[9]; and by extension, without a brain, there would have been no mind; and without mind, there would be no ideas, much less a "world" that could be labeled *ideal* and somehow separate from the physical. He continues, "It is not only the separation between mind and brain that is mythical: the separation between mind and body is probably just as fictional. The mind is *embodied*, in the full sense of the term, not just *embrained*" (118, italics added). From this perspective, the mental, spiritual, and metaphysical all derive from the physical in manners that Hamlet prototypically comes to sense in his own predicament. There are no two separate points of physical and nonphysical genesis; they are parts of the same world and in many significant ways are inseparable from each other, despite our linguistic and sophistic expostulations to the contrary.

Damasio goes a step further in *Looking for Spinoza: Joy, Sorrow, and the Feeling Brain* (2003) when he argues that "mental processes are grounded in the brain's mappings of the body," mappings that are "collections of neural patterns that portray responses to events that cause emotions and feelings."[10] He then observes, "[I]n a fruitful combination with past memories, imagination, and reasoning, feelings le[a]d to the emergence of foresight and the possibility of creating novel, nonstereotypical responses" (80). Put another way, mentality is not an isolatable activity; it physically requires input from the heretofore suspect physical environment by way of the body's conspiring senses and neural networking. A fundamental error in reasoning itself—initiated by and promoted in Western culture since at least the time of Plato or Socrates—has valorized the notion that inviolable truth can only derive from disembodied reason. Damasio is articulating in scientific terms what comedy has known all along: knowing requires feeling, and the body's outreach to its environment is central and instrumental to knowledge. Subjective engagement with our surroundings provides key ingredients to understanding, and that engagement undermines the presumed priority of isolated, rational, objectified mentality.

From this integrative perspective, it becomes important to tap into our "nature." We can all quickly recognize, for example, our inherent aggressive qualities; but rather than to rationally deduce that aggression entails retribution and death, perhaps it is worthwhile to adapt ourselves—self-consciously—to follow nature's tendencies to redirect aggression to less-than-mortal alternatives. It's not hardwired into our behavior to do so since humans actually outdistanced evolution through its cultural advances in weapons technology. But it's not beyond our ingenuity to draw from nature's lessons. This struggle between the encultured ideal of "honorable" aggression and the natural but dishonorable strategy of redirected aggression is precisely what Hamlet has difficulty resolving. Recall that Hamlet is a master of redirected aggression, beating everyone he encounters in games of wit and genius, outfoxing even self-professed foxes at every turn. It's a strategy that will backfire toward play's end when he agrees to aggressive redirection by agreeing to a game of swordsmanship with Laertes.

From this point of view, Hamlet's persistent deferrals are actually manifestations of his idealized vision of behavior adjusting to and aligning itself with life-prolonging natural habit. When Hamlet does act destructively, it is only when he reduces his human peers to a subspecies of creation, calling Polonius "a rat" and reducing Rosencrantz and Guildenstern to mere pawns, processes of sub- or "pseudo-speciation," according to Meeker, that allow us to look on some of our kind as "fair game" to be rubbed out as so many pestering flies about our heads.

This gets us back to the matter of seeming and being as we come to realize the interplay of mind and all the rest of the world beyond or other than mind. It doesn't take much to recall that consciousness actually introduced deception into our world at its inception since consciousness allows us to pursue what Damasio calls "nonstereotypical responses" contrary to our beings' simpler impulses, unqualified aggression certainly included. With consciousness, we can see and even control our behavior, leading to the common presumption that we can manipulate our behavior to *appear* contrary to our true feelings/emotions. However, behavioral scientists are beginning more accurately to understand that there are uncontrollable signals in human behavior that undermine this presumptive capacity to deceive, including that capacity for self-deception. In *The Feeling of What Happens* (1999), Damasio observes that "voluntary mimicking of expressions of emotion is easily detected as fake—something always fails, whether in the configuration of the facial muscles or in the tone of voice."[11] In other words, there is a pre-aware level of consciousness that triggers physical behavior, which is virtually impossible for humans to control and manipulate. Truth is in the body, and knowing the body requires enrolling the senses.

Damasio, a neuroscientist, pays particular attention to actors, masters of manipulation apparently capable of controlling their behavior in ways the rest of us can only admire. Hamlet, too, is intrigued by actors and their craft, especially those committed to anti-theatrical, realist *illusion* on the stage. For the early Hamlet, the stage is an additionally corrupt mirror of the real world of inherent deception. And it plays a part in Hamlet's own ventures to enter a world of dissembling while trying to avoid its entangling clutches, given the success at which Hamlet himself evolves into a consummate actor who leaves most of us standing before him in mute amazement, providing us with the supreme test of our abilities to read the body as revealer of truth. It truly is difficult to uncover deception among masters of deceit. But the fact that deception is possible and also difficult to uncover does not necessarily force us to abandon the physical world and turn to some parallel universe of disembodied ideas in our pursuit of truth. In our world, words often do fly up, and thoughts often do remain behind; and harlots' cheeks can be plastered over with seeming virtue. But we need not dispense with evidence from the physical world and conclude it to be mere inauthentic/disenchanted simulacra. Rather, our search for truth should instead redouble its efforts to strike through masks of simulacral deception in the world around us. This approach converts the duplicitous *teatrum mundi* motif to one that actually argues the indispensable nature of the seeing "world as stage," enchanted by images filled with meaning, mystery, and even truth. As Damasio insists, we can't think—we actually can't *be*—without engaging ourselves with that highly theatrical world "out there," full of far more talking bodies—if we learn to listen—than the talking heads we've come so foolishly to trust.

Using our senses and relying on the various mechanisms of contact with the world around us better positions us to distinguish inauthentic seeming from authentic being. Consider that Hamlet's own mousetrap experiment is an early-modern proof of Damasio's assertion: "The play's the thing" wherein Hamlet will "catch the conscience of the king" (II.ii.590–91) by having Claudius' guilty body proclaim its "malefactions" (II.ii.578). Consciousness can conceal the body's truth only so far. And what is true of Claudius' flawed efforts to deceive all of Denmark is also true ultimately even of Hamlet's efforts. Claudius is proof that bodies will confess truth despite mind's best efforts at concealment. What needs to happen is learning to read the body for signs of authentic action and behavior amid the myriad inauthentic diversions placed before us. Lowering himself to employing the theatre to catch the conscience of the king, Hamlet has before him the whole mechanism whereby he could catch the conscience of the world at large. But he's not at the point where he can see the truth of this point even as it perches itself before his very eyes.

This is a point that Hamlet does learn, but only after torturously pursuing myriad dead ends that lead to further fears of being deceived by simulacral inauthenticity. It clearly takes a while to occur; Hamlet begins to reveal his change in perspective following his aborted trip to England when he returns to Elsinore, a *body* literally affected by its physically inescapable attraction to Denmark against the will of Hamlet's idealist, escapist desires. His failed attempt physically to will his body to leave the gravitational attraction of the all-too-physical Elsinore marks the point where Hamlet intellectually becomes a different man, one who is awakened to seeing the physical world as more gravely real, but also as more authentic and enchanted, than before. More specifically, until this signal event, this literal sea change, Hamlet had virtually completely disregarded the significance of the physical realities of the world about him, hardly even grieving over the irreversible physical destruction of his father and Polonius, nor feeling any remorse over the cruelty of his behavior toward those around him. Recall that Hamlet couldn't care less for the body of the murdered Polonius but is now (in act V) fully prepared to grieve over the body of Ophelia. Bodies in all their unseemly appearance now matter. Until his return from the sea, he grieves over the loss of ideas/ideals like honor and virtue (exacerbated by his father's perfunctory funeral and mother's hasty marriage), but he has no real feelings for mere things of the flesh. This all changes after Hamlet's return to Denmark.

"There is a special providence in the fall of a sparrow" (V.ii.207), Hamlet concedes in the latter stages of the play, followed immediately by, "If it be now 'tis not to come; if it be not to come, it will be now; if it be not now; yet it will come. The readiness is all" (V.ii.208–11). These words could very easily be spoken by a comic protagonist like a Benedick (*Much Ado About Nothing*) or even a Bottom (*A Midsummer Night's Dream*), fittingly comic as they are in their concession that one must accept the often twisted and confusing ways of the world. As such, these words appear entirely inappropriate for an idealistic tragic hero, certainly one held in as high regard as Hamlet. Hamlet's singular resignation, fatalism, and acceptance of these fifth-act words seem at first glance to reveal nothing less than a total mental capitulation of a man who in the opening act seems so determined to take the world in his hands and restructure, indeed remake, that world as *he* sees fit. That initial Titan of power and willful self-confidence becomes someone, by act V, entirely inappropriate for a tragic frame (or for any traditionally heroic gesture whatsoever, for that matter). But by play's end, Hamlet is a mere shadow of this former grand (but self-deluded) self. A man who seemed to be a colossus looming over his domain is reduced to humbly comparing his lot to that of a lowly sparrow. So is this, as Ophelia prematurely notes in act III, the tale of "a noble mind . . . o'erthrown" (III.i.150)? Will, by play's end, we see little more than a "noble and sweet reason, like sweet bells jangled, out of time and harsh, blasted with [the] ecstasy [of madness]" (III.i.157–60)?

Hamlet's trajectory centrally involves a precipitous rise and subsequent revision of a dualistically fabricated Renaissance mind that was once dedicated exclusively to idealism over realism. The early Hamlet seems to *know* the truth that the world is rotten even before the ghost appears. His "prophetic soul" and all the other less-than-concrete evidence convinces him of Claudius' crime, yet he chooses to wait for concrete evidence. If Hamlet, a man who abhors this "too, too, solid" world, is actually committed to that abhorrence, then material evidence should have been irrelevant. But he's not even sure whether "to be" or "not to be" is the preferred condition. Surely, these are not words of a committed idealist. Surely, the casually suicidal Socrates would never have spoken them. Ultimately, through the fog of apparent madness, Hamlet's mind wraps around an integrationist idea that the ideal and physical dissolve as concepts in ways that were undreamt of in his previous philosophy. His epiphany converts him from an inflexible tragic visionary to a changeable, Protean, even comic presence ready to adapt to material circumstances and hungry for the life that the new vision of redirected aggression tantalizingly places before him.

But it's a vision difficult to actualize, as Hector's capitulation and Hamlet's final moments among the living demonstrate. First, Hamlet muses quietly on the nature of death ("Alas, poor Yorick" [V.i.183]) with a poignancy not found in the early mannerist "to be or not to be" soliloquy. Then he proclaims something that has not been spoken throughout the play—and with a passion and presumed honesty not heard before, either:

> I loved Ophelia. Forty thousand brothers
> Could not with all their quantity of love
> Make up my sum. (V.i.272–74)

Emotion is allowed and so are attendant fleshly thoughts on love and mortality. Hamlet's final motivation to kill Claudius seems to be his preservationist discovery that Claudius had tried to kill Hamlet himself and would likely try again. Hamlet has come to realize now his newfound unwillingness to give up his life on earth, that thing he once valued at less than a pin's fee. But it also seems that the ideal of revenge has lost sway altogether in Hamlet's mind, if it ever really was part of his plan.

Amid all this engagement in the events of the mortal physical world, Hamlet confronts Laertes with a plea for pardon:

Give me your pardon, sir. I have done you wrong,
But pardon't as you are a gentlemen.
This presence knows,
And you must needs have heard, how I am punished
With a sore distraction. (V.ii.224–28)

He goes so far as to proclaim his past actions "madness" (V.ii.230) and argues with some accuracy that this madness has been "poor Hamlet's enemy" (V.ii.237). Yielding to any man, even to a man Hamlet has so wronged, is clearly not an action imaginable in an earlier Hamlet. But the justice Hamlet seeks at this late stage involves a confession of personal complicity in a world now run riot in part because of his previous refusal to properly accept his place in the world.

It is Hamlet's hope that retribution is no longer the force of nature that it once appeared to be. For Hamlet, it has become a troubling delusion against which humanity must struggle, either by the redirection he earlier utilized or by actual confession of personal guilt which he now tries. Aggression, it seems, has reached its limit in Hamlet's agenda. Laertes properly understands and nearly accepts Hamlet's request for peace among aggressors. But once again we see deception loosed on the world as Laertes tells Hamlet, "I do receive your offered love like love" (V.ii.249) only to mortally wound Hamlet even as Laertes confesses, "[I]t is almost against my conscience" (V.ii.298). In his enthusiasm, Hamlet has forgotten that his personal transformation to a new vision of human behavior is a singularity within this world still governed by retributive aggression and attendant duplicity of seeming virtue. He fails to recognize that Laertes—much wronged—and the world at large have much further to go to reach Hamlet's new state of mind and being.

The world, finally, is not prepared to accept the fruits of Hamlet's late-arriving epiphany. It is a transformative vision that alters the very foundations of ontology and epistemology. Knowing what it is and understanding how one knows it are clearly key breakthroughs for Hamlet. And attached to these revelations comes a curious ethical component. The aggression that attended Hamlet's struggle to maintain autonomy amid a physical world bent on swallowing him up somehow dissipates, transforming into a recognition that aggression may be humanly inherent but that redirection is a naturally endorsed path that could lead to increasing the odds of survivability. In fact, Hamlet sees the physical interrelatedness of a grand design that includes even the fall of a sparrow. To struggle against this grand design, he now realizes, is the height of human folly, a comic revelation if ever there were one. It comes too late to save Hamlet—or Elsinore—much as it remains tardy in our own world.

This gets us back to Segal's observation, quoted earlier: "The tragic hero dies for what is nobler in the mind, the comic hero *lives* for what is humbler in the flesh" (12). So much of what Hamlet becomes by play's end makes him a comic rather than a tragic hero. By the time he says, "There's a divinity that shapes our ends" (V.ii.10), Hamlet has in fact "humbled" himself to the life of living in the flesh. But the bloody events of the world have gone too far to be pulled out by bloodless comic resolution-by-redirection. Comedy struggles for a foothold, but tragedy holds the stage. Comedy and tragedy on either side of a razor's edge—tragedy wins this round, returning order to Elsinore, purchased with blood that need not have been spilled for order to return. And the illusive promise of a *new* order flickers for a moment but then recedes as those fortunates who avoided the carnage clean up the stage and brace for future aggression and retribution.

CHAPTER 3

Connecting Mind to Body

Hamlet moves from his idealist perch to the too, too solid (and even sullied) world in which he inescapably belongs. Coming to understand that there are consequences to being bounded by the world around us, Hamlet experiences his act V transformation unfortunately too late to save himself and Elsinore from an unnecessary bloodbath. Hector sees something of the same point but is convinced to act upon old ideals. Troilus never approaches understanding his place in the world and so miserably holds to his high-mindedness as his world spirals to an infernal doom.

While Hamlet, Hector, and Troilus cannot ignore the mortal fact that they were physically born into this world and are therefore part of the world, the idealist in them argues that we should quickly come to realize the world's imperfect nature, turn away from it, and strive for the ideal. That ideal, of course, inheres in ideas, of which the physically real is at best a mere shadow. And those ideas either inhere in or can be accessed through the mind alone. For this system to make sense, of course, one must presume that the body and some disembodied mind/spirit/soul are separate realities, a thesis relatively easily accepted if we tap into our common-sense experience that "me," "my thoughts," and "my body" are clearly distinct from "the rest of the world."

As Hamlet comes to realize, however, a problem arises when one subscribes to dualism, which is then compounded when one chooses (as one must) to privilege one side over the other: things just don't add up. And it's not a matter of having chosen the wrong side. Rather, it's a matter of presuming there's actually even a side to take. Writing from a twentieth-century perspective, neuroscientist Antonio Damasio reminds us of the early modern arrival of Cartesian dualism, but then observes, "Of late, the concept of mind has moved from the ethereal nowhere place it occupied in the seventeenth century to its current residence in or around the brain."[1] He adds, "It is not only the separation between mind and brain that is mythical: the separation between mind and body is probably just as fictional. The mind is embodied, in the full sense of the term, not just embrained" (118). From this point, Damasio does concede that this point is "a bit of a demotion, but still a dignified station" (225) for humanity to occupy.

That the mind is embodied has today become a *scientific* given, but that doesn't mean that we necessarily accept the point as a *cultural* given. It appears that the demotion is and has been harder to overcome than would be presumed, given the facts before us. Consider our attitudes toward mind and body/environment even to this day. We remain obsessed with taking control of these things called our bodies, looking at them as objects that tend to work against our better natures to drag us into temptations that will lead to our early demise. And the natural world about us must also be cleaned up lest we risk contact with its changeable essence and fall further victim to the spatio-temporal non-ideal. Put it in order, make it conform to ideals derived from mind, and make it serve our restless ambition to make this fallible physicality more like the mental ideal we know to exist beyond this irregular patch of materiality.

Consider not only the damage to nature that this us-versus-it perspective has generated, but also the psychological damage stemming from such posturing. Being clearly separate from that which physically surrounds us leads to the glorifying notion of individual/personal autonomy. But, rather paradoxically, the price of this individuation is individuation itself, for in order to achieve unencumbered autonomy, we must find a way to cast off the results of our ennaturalizing childbirths. So the "natural" urge of our bodies to remain ennatured—often characterized as a longing for continued bonds with the maternal—must be overcome by the distancing lessons of a stern cultural/paternal authority that demonizes the natural and valorizes the ideal. There's no need here to go into the myriad competing theories concerning the proper socialization processes for human beings. What is generally true for virtually all of them is the presumption that the separation from nature is good, correct, necessary, or humanly inevitable. And the result of these presumptions and practices is that, in order to maintain at least an illusion of individuated autonomy, humanity has turned to warring against a variety of threatening "others," including other humans, an othered environment, and even itself as other. The very possibility that somehow we should *not* be fighting against the truth that we are ennatured/environmented beings seems to have escaped our culture's consciousness in large measure.

So the notion that mind exists in an ethereal nowhere place is far more than a matter of mere philosophical debate. The ascendancy of Cartesian dualism in Western thought and its related dualist antecedents (of which

Platonism and Gnosticism are two examples) have occupied a primary position in intellectual circles and high European cultural orthodoxy (including Roman Catholic Christianity) for centuries. And the consequences can be said to be both liberating (generating a human valorization of individualism) and debilitating (generating human byproducts of warlike aggression and despairing solipsism). But there has also lingered a relatively strong undercurrent that operates in contrast to that orthodoxy, found in less aristocratic or clerical locales and hence easy to be overlooked. The body did not wholly lose its place in Western ontology and was even celebrated on rude, vulgar, common occasions like carnival and market-day festivities. Highly suspect entertainments celebrated primarily by culture's underclasses looked and smelled suspiciously like a celebration of the physical over the mental, requiring vigilance on the part of those controlling cultural forces that were opponents of the flesh (a point excellently brought home in Jonas Barish's *The Antitheatrical Prejudice*[2]).

This countercultural urge toward viewing the body as a partner in selfhood of course precedes twentieth-century breakthroughs in neurobiology. One fruitful Western antecedent is medieval humorology, a point made by Mary Thomas Crane in *Shakespeare's Brain* (2001), who observes that the current "cognitive concept of an embodied mind seems closer to early modern humoral physiology than the radically dualistic post-Cartesian paradigm."[3] In other words, a strong theory *interconnecting* mind and body was active throughout and even prior to the Elizabethan age, apparently coexisting with high culture's idealism. Humorology was one tool usefully utilized as early modern playwrights pursued the comic agenda of embodying selfhood. What we see are parallels that compare well to our continuing, contemporary cultural pursuit of the same.

Humors Uniting Mind and Body

David Hillman, in "Visceral Knowledge: Shakespeare, Skepticism, and the Interior of the Early Modern Body" (1997), observes that "selfhood and materiality . . . were ineluctably linked in pre-Cartesian belief systems of the period, which preceded, for the most part, any attempt to separate the vocabulary of medical and humoral physiology from that of individual psychology"[4] The early modern institutionalization of the idea of human connectivity to the world about us is brilliant in its simplicity and at least metaphorically in line with twenty-first-century conceptions of mind and body.

The body absorbs food and converts it into four liquid substances called the humors—melancholy, phlegm, blood, choler—each of which has its own counterpart among the elements—earth, water, air, fire. To be physically and *psychically* healthy, an individual must consume a balance of elements and convert an attendant balance in humors. Exactly how the body and mind interact with the world was further theorized within the system. E. M. W. Tillyard summarizes the tripartite system: "Like the body, the brain was divided into a triple hierarchy. The lowest contained the five senses. The middle contained first the common sense, which received and summarized the reports of the five senses, second the fancy, and third the memory. This middle supplied the materials for the highest to work on. The highest contained the supreme human faculty the reason, by which man is separated from the beasts and allied to God and the angels, with its two parts, the understanding (or wit) and the will."[5] The world supplies material for the reason to work on even as that reason is physically nurtured by the world it mentally engages. Mind and body are functionally inseparable, and no amount of skepticism regarding the validity/accuracy/truth regarding sensory input can separate the two merely *apparently separate* realms. "Man is conspicuous," says Tillyard, "by seeking perfection through knowledge external to himself" (79), meaning, "[t]he understanding . . . had to sift the evidence of the senses already organized by the common sense, to examine the exuberant creations of the fancy, to summon up the right material from the memory, and on its own account to lay up the greatest possible store of knowledge and wisdom" (80). Humanity, in short, is the sum of all that it consumes, orally, sensually, and otherwise. What is spiritual in us is not so much a case of "god" breathing life/spirit into us from above, but of base physical materials being refined in the body and "rising" to the realm of the mental/spiritual. The idea implicit in humoral physiology includes a sense of human interconnectivity with the world and others about us, even as traditional Judeo-Christian religion paradoxically continued its call to otherworldly regard. So goes the medieval world picture.

Then, in a strange Renaissance twist, otherworldly regard eventually and rather ironically received secular endorsement as Renaissance humanism developed a sense of individuated autonomy and chipped away at any sense of a need to prop up the "new man" with external physical scaffolding. This growing sense of self-reliance combined with a renewed philosophical skepticism in the reliability of the human senses to return the flesh to the world of shadows and suspicion. If humanity were able to stand triumphantly alone in the world—as the Renaissance claimed it could—there was no real need to search for external/physical causes for human behavior nor to succumb to the humbling pleasures of the senses (though manipulation and indulgence of the senses became, more than ever, singular temptations). As autonomous creatures, surely we were masters of our behavior. Such a growing sense of human autonomy led to an increased sense of humans' dominion of and separation from the world around us. And the sense of separation opened the door to the virtually complete intellectual triumph of what became Cartesian dualism.

We may have here a reason why tragedy remained basically dormant throughout the Middle Ages and then arose with a vengeance in the Renaissance. The ascendancy of post-medieval idealistic perfectionism led to increased recognition of and respect for tragedy as it revealed the humanist idealist's conclusion that the human condition is a tragic condition involving (nearly) impossible aspiration. Wanting to disconnect ourselves from a once much-needed materially explanatory scaffolding, we became reacquainted with tragic independence from and resistance to externality.

Comedy, however, reminds us at very least of our first needs—namely to establish a physical "humors" balance within ourselves. Ben Jonson (ca. 1572–1637) was, of course, the master of humors on the English Renaissance stage, a man who, according to Una Ellis-Fermor, treated his material in a way that could be called "scientific for its endeavour to present moral and psychological truth more and more nearly in terms of actuality and to eliminate more and more thoroughly the subjective."[6] Engaging the humors tradition as he does, Jonson's "science" draws from medieval humors, employs the related ancient Greek concept of *hybris*, and adapts the modified humors conventions of his own age. *Hybris*, as Helen Ostovich notes, "denoted the aggressive spirit of display in certain animals, like the strutting and crowing of cocks."[7] She continues, "In humans, it appears as immoderate self-absorption: an excessive spirit of aggression fostering excessive indulgence which then fosters the excessive aggressiveness, *ad infinitum*. It may express itself as excessive eating, drinking, or sexual activity . . . or it may apply to any venture which expends surplus energy wastefully or ludicrously" (14). Ostovish further observes of humors itself, "Throughout the sixteenth century, the term had acquired transferred meanings related to disposition and by the 1590s frequently appeared in studies of manners to suggest whim, affectation, dominant mood, or ruling folly" (13). Having lost some of its scientific/physiological purity, Jonson's "humors plays" nevertheless reflect a sense that folly is a result of humorous imbalance resulting from imbalanced overabsorption of certain embodied environmental factors. But since folly is connected to materialist influences and is therefore subject to correction, we are not necessarily inescapably or inevitably doomed by inherent, innate, or genetic defect. In other words, the "mind" has not arrived in the world so much in a defective condition as it has been poorly nurtured by the environment into which it has been placed. Its embodied reality literally affects its abstracted musings, if ever there were true abstractions as we often conceive of them.

The consequence of this humoral theory is that human defect is seen as correctible through exposure to balancing environmental influences. It is, as Bergson would say, the height of human vanity to foster excess and the aggression that excess generates. Returning to a "natural," embodied balance is the goal of the human organism, not the urge to rocket beyond the gravitational influence of our embodied beings. From this perspective, the very idea of abstract thought could be conceived of as a sort of madness stemming from isolation generating a mind mired in abstracted fantasy rather than connected to reality. More connected to materiality and therefore less potentially tragic is the idea of human folly resulting from a humorous, embodied imbalance that is correctible through proper exposure to elements of rebalance. Tragedy, in other words, derives from a willfully arrogant refusal comically to accept material correction.

Jonson's *Every Man Out of His Humour* (1600) is such a work designed to convey the folly of human, materially generated humors imbalance. The text of the play introduces each character with a brief labeling sketch: Asper is "of an ingenious and free spirit," Puntavarlo a "vainglorious knight," Carlo Buffone, a "public scurrilous, and profane jester," and so on. What Jonson does in his play with his cast of humors is twofold: (1) he offers correctives in the most extreme cases of imbalance, and (2) he demonstrates the human need for adaptive flexibility in dealing with and living in one's environment.

The first strategy is an obvious satiric strategy. Asper, the play's inveterate critic of human folly, outlines his intention in the Induction. Warned by Mitis that "men are grown impatient of reproof" (Ind.123[8]), Asper responds that they are

> None but a sort of fools, so sick in taste
> That they contemn all physic of the mind,
> And like galled camels kick at every touch. (Ind.130–32)

In contrast to these resistant fools, Asper observes

> Good men and virtuous spirits that loathe their vices
> Will cherish my free labours, love my lines,
> And with fervour of their shining grace
> Make my brain fruitful to bring forth more objects
> Worthy of their serious intentive eyes. (Ind.133–37)

Fools—of which tragic heroes may be a subset—resist correction, while good men and virtuous spirits seek correction. For Asper, however, apparently *everyone* needs correction, his assaults on his fellows being "medicinally" intended to cure people of their humorous follies. Extending his medical metaphor, Asper observes that the fool cloaked in the garb of respectability and therefore subject to emulation is as dangerous as a carrier of an actual disease because, in London at least,

> They're more infectious than the pestilence,
> And therefore I would give them pills to purge,
> And make 'em fit for fair societies. (Ind.173–75)

The pill, of course, is satiric exposure to ridicule, a sure humors cure to folly. Strong medicine is needed for powerful ailments, and those requiring physic must be appropriately convinced of the need, through badgering or humiliation, perhaps, or even possible coercion of other forms. At least that's how Asper sees the matter.

Connected to this missionary zeal, Asper's view has a notable limitation, given that he singularly believes there is, somehow, a single "right" action/behavior. This, in fact, is something of a singular problem with the satiric tradition in general, from Aristotle to Bergson and beyond. The general presumption of the existence of a universal social "norm" is a presumption that is, at best, highly suspect. It is not a suspicion affecting Jonson's Asper, however—at least not initially. He believes that physicking all patients to arrive at a singular construction of self will result in a healthy reintegration into and strengthening of a harmonious social fabric. Manufacture a single "proper" humorous balance and all will be well. However, as Ostovich observes, "This would be to minimize the varieties of life that emerge in rascals like Carlo or Shift, or in courtiers like the stiff-limbed Puntavarlo or the transparent poseur, Fastidious Brisk. Rather, the whole spectrum of roles, actors and audiences can enrich the understanding of social performance by showing that a single model of behaviour is inadequate in a social process that demands a complex shifting from one role to another" (83). Replacing Asper's essentially unilateral and inflexible single-mindedness, Jonson acknowledges a wider perspective. Variety among human behavior seems the additional condition Jonson advocates. Balance is crucial; but uniformity, especially mandated uniformity, is to be devoutly avoided. Ostovich concludes of the play's conclusion, "With all humours now written and played out, Jonson does not mean to offer his spectators a narrow 'humourous' judgement as a final point of view, but to let them alone to 'feed their understanding parts' (Ind.201)" (83). The singular satiric physic of Asper and his like inaccurately advocates a one-size-fits-all behaviorism, an inflexible *materialistic* brand of idealism not altogether different from what surfaces in the nonmaterialistic tragic hero. Personality construction, finally, requires the flexibility to adapt to a multiplicity of given circumstances, a multiplicity of environmental factors. The humors must adjust according to conditions into which the body, brain, mind, and self is thrust.

Stern satire often ignores the importance of this interactive notion of selfhood formulation. Virtue frequently requires humorous adjustments; but if there is a prescription for correction that disregards the patient's environment, then we have once again moved onto dangerous terrain. Satire in this case moves closer to tragedy than comedy, disconnected as this brand of comedy has become of its material roots. One in fact can look at Hamlet himself—the early Hamlet, that is—as a consummate and consummately inflexible satirist, suffering fools poorly and even killing them off without any apparent remorse. Giving up that role, Hamlet eventually moves a major step away from his melancholic, inflexible, and therefore tragic humor and toward a comic sense of personhood willing to adjust to the world around him. Though Hamlet's conversion wasn't too little, it was clearly too late, at least too late to secure Hamlet's survival, though his transformation does contribute to offer, in part, the necessary physic for his sicklied o'er Denmark. It's the sort of adaptive humorous revelation Jonson hoped to see wash over his own London audience, preferably in time to avoid the unsavory fate of Hamlet's Denmark.

One can only wonder what sort of world would have evolved had Asper or the early Hamlet actually had their visions realized. (The world that Oliver Cromwell and the Puritans actually tried to create may have been that world.) In the same vein, one wonders at the misanthropic, plain-dealing Alceste created by Molière (1622–73). In the introduction to his translation of *The Misanthrope*, Richard Wilbur observes, "In this play, society itself is indicted, and though Alceste's criticisms are indiscriminate, they are not unjustified. It is true that falseness and intrigue are everywhere on view; the conventions enforce a routine dishonesty, justice is subverted by influence, love is overwhelmed by calculation, and these things are accepted, even by the best, as 'natural.'"[9]

He adds, however, "But *The Misanthrope* is not only a critique of society; it is also a study of impurity of motive in a critic of society. If Alceste has a rage for the genuine, and he truly has, it is unfortunately compromised and exploited by his vast, unconscious egotism. . . . Like many humorless and indignant people, he is hard on everybody but himself, and does not perceive it when he fails his own ideals" (7). Alceste, Asper, Hamlet—and Troilus, too—succumb to personal and inflexible vanity as a result of their hot pursuit of their ideals.

What should be added here is that comedy's "norms," when asserted in the best of comedies, rarely truly or effectively espouse moral or ethical norms without in the process espousing something significantly different than comedy typically espouses. An audience, for example, would be "correct" to identify with Alceste's misanthropic posture if *The Misanthrope* were a moral treatise. And we do sympathize with Alceste, to a point at least. But then we're hit with a sense of revulsion at Alceste's arrogance, leading us to another almost contradictory "feeling" that something is more "wrong" with him than with the world. Here we're reminded of Bergson's observation that a "flexible vice may not be so easy to ridicule as a rigid virtue. It is *rigidity* that society eyes with suspicion. Consequently, it is the rigidity of Alceste that makes us laugh, though here rigidity stands for honesty."[10] Alceste's inflexible unwillingness to negotiate with and to adapt to the worldly standards about him becomes his flaw. Even though his moralistic stance is formally "correct," civility dependent on occasional white lies is a necessary glue for human sociality and therefore trumps moralistically precise plain-dealing as a dominant virtue. For comedy, social adaptability is key, not idealized morality. By giving Alceste his say, however, Molière creates an ingenious work in *The Misanthrope* that satisfies moralists even as the play as a whole serves the larger comic spirit. So the point that even moralistically accurate satire can be satirized returns us to the idea that a desire to correct an imbalance can lead to an imbalance of perhaps even greater consequence. When the cure is a remedy inflexibly applied, it becomes worse than the ailment it was intended to remedy.

As such, in a world of people possessed of humorous imbalances requiring necessary physic adjustment, the shortcoming of critical intolerance found in the likes of Asper, Alceste, and others ironically stands out as *the* failing of all failings. If others are directed to adjust to the worlds around them, and if in the process others are to influence an adjustment to the societies they occupy, these proud idealists are themselves in need of even greater adjustments than their "patients," primarily because they choose to try to stand outside the world of which they are nonetheless part. But it's not so much the case that they must conform to a norm that they abhor. Rather, they should develop something akin to what Susan Purdie calls "a discourse of exchange,"[11] involving themselves with the world around them and hoping to exact adjustments, compromises, and adaptations suitable for all concerned. Refusing the humbled position of being part of their world, they stand in judgment over it, little realizing that this very posturing is the result of serious—sometimes even fatal—misjudgment of their true place. Troilus stands alone and apart in his world; Hamlet departs, escorted by flights of angels; and Alceste the misanthrope leaves civilized society, mock-heroically blessing his erstwhile friends:

> May you be true to all you now profess,
> And so deserve unending happiness.

> Meanwhile, betrayed and wronged in everything,
> I'll flee this bitter world where vice is king,
> And seek some spot unpeopled and apart
> Where I'll be free to have an honest heart. (V.viii)

Though his friends follow after him in hopes of changing his mind, it is his "honest heart" that first needs physic and then perhaps through that change his mind will change, too. A true change of humor, a need to connect to the material world rather than adhering to an abstracted ideal will need to occur for Alceste truly to remain on this stage.

By play's end, Asper seems to have begun to learn such a lesson, ending *Every Man Out of His Humour* placed before a censorious audience to whom he confesses a personal change of attitude: "I stand wholly to your kind approbation, and, indeed, am nothing so peremptory as I was in the beginning" (V.iv.57–58). He ends, "[I]f you, out of the bounty of your good liking, will bestow it, why, you may, in time, make lean [Asper] as fat as Sir John Falstaff" (V.iv.61–63). Choosing life and full, humorous, material engagement over the humorous failing of idealistic, abstract censure, Aster seems on his way to recovery.

Banish Falstaff and Banish the World

If one is looking for a dedicated, unrepentantly anti-idealist corporealist, Shakespeare's Falstaff is the man. Though planted in Shakespeare's history cycle of plays, this carnally indulgent mound of flesh so controls his scenes that even his banishment (in *2 Henry IV*) and subsequent death (reported in *Henry V*) do little to diminish his lasting comic impression on the stage. In fact, when Prince Hal assumes the role of king in *Henry V*, ruling his domain without Falstaff anywhere near, Shakespeare is left with a much diminished play, having created a dull world lacking any viable foil against which Hal can match wits. *Henry V* actually becomes a rather mechanical product involving little more than statically heroic, idealistically encrusted myth-making manners, form, and routine. No idealist/heroic/chivalric/tragic Hotspur of *1 Henry IV* arises, only a band of ineffectual French nobles. And in place of corporeal Falstaff is everyone else, hopelessly incapable of looking squarely at Hal and fulfilling any role other than easy capitulation to this historical giant. Order may be achieved but at the cost of living. Our minds may be relieved by the stability found in *Henry V*, but our bodies cry out for much more.

But there's something of an important Falstaff legacy that lingers in *Henry V* even after he's gone. First of all, there's some comfort in knowing that Hal-turned-Henry carries with him an understanding of his English underclassmen that allows him to rule his domain as no recent (or immediately future) king of England has been able. Falstaff's misrule on the earlier stage has infected Hal/Henry in no inconsiderable way, putting Hal through a fire of comic transformation that touched Hamlet too late, Hector to a point, and Troilus not at all. Then there's the added quality that Falstaff implicitly contributes to the even larger fabric of the culture he at least initially destabilizes. In *Homo Ludens: A Study of the Play Element in Culture* (1944), Johan Huizinga[12] articulates the point that humans are born with a virtually unique capacity to play, noting that "play" is of paramount importance to human social health. The point can be updated to include the notion of redirected aggression and how it functions to defuse the deadly potential in humanity to destroy itself when aggression lacks effective, nonlethal alternatives. Recall that *Henry V* opens with a clearly metatheatrical invocation that the play is actually a play. It could be argued that this invocation is a formal though implicit nod to the life-sustaining effects of Falstaff's grand theatricality. Put another way, that invocation acknowledges what Falstaff's earlier life on the stage demonstrated, that life itself is play and that—in direct conflict with Hamlet's idealistic insistence he can tell the difference—seeming actually is a crucial form of being. It's not just a matter of decrying seeming as being little more than vile deception to be identified, confronted, and overcome. Rather, it's a case that seeming, that putting on an antic disposition, is fundamental to human survival.

In fact, the very notion that there is a category of stable being within nature is called into question, a "natural" presumption or ideal that has no natural basis whatsoever. In *Madness, Masks, and Laughter: An Essay on Comedy* (1995), R. D. V. Glasgow echoes the point by observing, "Role theorists (amongst others) maintain that the idea of a private core within a person is a fiction: the inner, 'true' self, they claim, cannot be separated from the rest of our role-playing, for the human being is a social actor not by choice but by definition."[13] Even the most basic or natural of acts—he cites sexual intercourse—has a fundamental ludic quality. And in a statement that curiously unites both Hamlet and Falstaff, Glasgow observes that "pretending to be a liar is virtually indistinguishable from actually being one" (194), given that both lead to similar consequences. But while Hamlet continues the fruitless task of trying to distinguish the two, Falstaff's comic vitality occurs because he embraces his ludic self to the point that "he becomes his role, he lives his act, for his identity is essentially fluid" (195). This perspective, of course, explains the comic centrality of disguises, masks, and mistaken identity. But from a position of negotiated expectations, these forms of confusion do not so much need to be settled in order to find a character's "true" self; rather, they surface, and they are tested and altered to demonstrate the human ludic need for changeability itself, a willingness to negotiate and adapt to the changeable environment in which we are immersed.

So when Falstaff is present, life lights up with full vitality, especially within the playful deception he so masterfully generates. Though aware of Hal's rank and future, Falstaff takes on the role of Hal's master and even dons the garb of Hal's father, irreverently challenging Hal's authority at every turn and sharpening Hal's own ready wit in the process. Eating, drinking, and whoring are Falstaff's passions, all exercised to excess. But there is more to Falstaff's excess than meets the eye (or ear or nose). To succeed in these "professions" of the flesh as fully as Falstaff does requires a shrewd, manipulating mind grown out of the flesh, designed to serve the flesh as it undermines anything and anyone determined to explore the more "honorable" ideals, laws, and codes that may run contrary to the desires of the flesh. In short, his mind is called on to serve his body to a frighteningly high degree as that body escapes any slings and arrows—literal and metaphorical—cast its way, redirecting aggression by playing the fool at every turn. When cornered, he never falters but rather cloaks himself in rationalizations so baldly flabby that "ignoble" laughter replaces "noble" indignation, even to the point that the law itself forgets its charge.

There may additionally be a very legitimate and timely method to Falstaff's madness. W. H. Auden makes the case that the wars in which England was engaged during Falstaff's reign on stage are civil wars virtually meaningless to any but the power elite. This is no Norman conquest wherein England's survival is at stake, but only squabbles among English nobles and of no real significance when it comes to England's survival. As such, Falstaff's less-than-heroic behavior actually surfaces as a "comic criticism of the feudal ethic as typified by Hotspur." Auden continues, "Courage is a personal virtue, but military glory for military glory's sake can be a social evil; unreasonable and unjust wars create a paradox that the personal vice of cowardice can become a public virtue."[14] Perhaps one reason Falstaff disappears in *Henry V* is that the struggles described at this stage in Shakespeare's history cycle apply more directly to England's actual survival and the eventual prosperity of all. And so Falstaff's perhaps inadvertent but nonetheless "satiric" assault on English actions and customs becomes inappropriate for the time, warranting his exit from the stage. The civil wars found in the earlier *1* and *2 Henry IV*, on the other hand, only determine personal prosperity or doom for a small band of noble antagonists and so provide opportunity for both the satire and celebration that Falstaff provides.

Shakespeare adds further, nonsatiric value to the destructive, civil war–torn world of the *Henry IV* plays by making that world Hal's training ground. It is the place where Falstaff's exuberant celebration of life can prevail over the ideals and virtues that are of little *real* value, given that they are used for the limited advancement of extreme self-interest among the greed-ridden English nobility. Survival or preservation of a nation and people is not at stake so much as preservation of personal vanity, creature comfort, and private prosperity. Alongside these carefully orchestrated forms and trappings of the feuding lords, Falstaff's own indulgences don't seem so different. As such, Falstaff's behavior is more of a bare-boned, hypocrisy-free reflection of the hollow self-interested behavior of the dueling aristocrats, with the difference being that the latter cloak their behavior in chivalrous trappings and shed far more blood than Falstaff's actions. (Falstaff, it would seem, would have thrived in *Troilus and Cressida*.) Exposing Hal to such bare-boned realities surely leads to a rounder, sounder soon-to-be king.

But there's still more. Perhaps the biggest difference between the feuding nobles and Falstaff is precisely that Falstaff so unabashedly adopts a life "style" that triumphs over everything that could impede his pleasure, honor and death certainly included. Even when he puts on an antic disposition of deception, it is hard to conceive that Falstaff is really fooling himself in the same manner that the play's ideologues do. For example, while playing the role of Hal's father, King Henry IV, Falstaff describes himself as a "Goodly portly man, i' faith, and a corpulent; of a cheerful look, a pleasing eye, and a most noble carriage; and, as I think, his age some fifty, or, by 'r Lady, inclining to threescore. . . . If that man should be lewdly given, he deceiveth me; for, Harry, I see virtue in his looks. . . . [p]eremptorily I speak it, there is virtue in that Falstaff" (II.iv.335–41).

Having a good opinion of oneself is, of course, a vital survival quality. Falstaff's self-described cheerful appearance, pleasing eye, and noble carriage run contrary to Hal's description of the man: "This sanguine coward, this bed-presser, this horse-backbreaker, this huge hill of flesh" (II.iv.195–96). But the frank truth of Hal's assessment in no way results in Falstaff crumbling in the face of such descriptions because his confidence in himself overrides any self-doubt that such assaults should generate. Nor does Hal abandon his friend for being self-deceived, arrogant, or larger than truth. Rather, this huge hill of flesh, this truly master thespian, has such gravitational "attraction" that his buddy and other colleagues—and likely audience members as well—are hooked by his masterful play, his ability to deflect slings and arrows of all shapes and sizes, and his capacity to thrive in ways that none of his cohorts seem able, despite equivalent opportunity.

Ultimately, one begins to sense that there *is* a certain nobility and virtue inhering in this hill of flesh, though such nobility and virtue are clearly of different orders than traditionally posited, conformable not to social prescription but rather to what idealizing society would derisively label laws of nature. He is not physically attractive by any standard formulation, lacks loyalty to his friends and respect for his female acquaintances, holds no awe-inspiring rank or position, and refuses to stand on any point of martial honor whatsoever. He's a braggart warrior, a stage clown of radical humor, a medieval vice figure. Even honor among thieves, a curiously acceptable inversion of ideals, is lacking in Falstaff. There really is nothing that Shakespeare gives to Falstaff, except a consummate ability to indulge his hunger for life, that stands in complete aversion to spending even a moment trying legitimately to earn a livelihood based on anything as tedious as personal fortitude, worth, or hard work. He is what he is because he is living flesh finding its way through the world by assuming a flexibility in life that even Hamlet would have had to admire.

That living flesh, served by a nimble mind, thrives by its wits. It is, of course, imaginable that such a lusty character could survive through simple intuitive empty-headedness. After all, simpletons abound in Shakespeare and elsewhere—consider Dogberry and Verges—whose successes are primarily matters of dumb luck. Others survive by simply hiding in shadows and corners, timidly reaching out for scraps left behind after great actions and grand ceremonies. But these are not Falstaff's ways. Living by his considerable wits, he takes center stage, accepting whatever ridicule being on center stage he is exposed to but surviving and even profiting from the exposure through the adept facility of his mind. For Falstaff, the mind serves the body, and his mind is no small matter. That is Falstaff's triumph. His mind has been cultivated to accumulate vast stores of ready information and "noble" allusions, but Falstaff utilizes that cultivation purely for the purpose of securing more life. As such, he is not a "natural" fool, a buffoon living by dumb luck soon to be pushed aside and left in the gutter when dumb luck runs out. Attention must be paid to this man, whose not inconsiderable capacities have been converted to serving the body, which houses the mind that develops and manipulates those capacities that serve the body. Falstaff is a fine-tuned survival machine.

He may be no king or heir apparent, but he can nevertheless stand toe to toe with Hal on the field of wit and insult. For example, following Hal's "huge hill of flesh" assault, Falstaff retaliates with, "'Sblood, you starveling, you eel-skin, you dried neat's tongue, you bull's pizzle, you stock fish! O, for breath to utter what is like thee" (II.iv.197–98). For what it's worth, these aggressive assaults are actually terms of endearment. Everyone can attack Falstaff in such a manner and without fear of mortal reprisal because he takes the assaults in the spirit that he himself utters them in response: altogether in a manner of play and at most a healthy manifestation of redirected aggression. He is masterful, too, in his own defense: "If to be old and merry be a sin, then many an old host that I know is damned. If to be fat is to be hated, then Pharaoh's lean kine are to be loved. No, my good lord, banish Peto, banish Bardolph, banish Poins; but for sweet Jack Falstaff, king Jack Falstaff, true Jack Falstaff, valiant Jack Falstaff, and therefore more valiant being as he is old Jack Falstaff, banish not him thy Harry's company, banish not him thy Harry's company—banish plump Jack, and banish all the world" (II.iv.375–82).

Hal, of course, will eventually banish Jack Falstaff, but not until Hal first sucks sufficient life from this vital organism. Hal learns from Falstaff the power of wit and the effect of rhetorical flourishes. Note, for example, Falstaff's classically structured/balanced defense above. Falstaff is no mere "natural," if such a thing even exists; rather, he has that within which embodies show in all its positive, celebratory affect. Recall, too, how Falstaff defends his behavior at Gad's Hill when the attempted robbery is foiled by the disguised Hal: "Why, hear you, my masters, was it for me to kill the heir apparent? Should I turn upon the true prince? Why, thou knowest I am as valiant as Hercules, but beware instinct. The lion will not touch the true prince. Instinct is a great matter; I was now a coward on instinct. I shall think better of myself and thee during my life—I for a valiant lion, and thou for a true prince" (II.iv.213–18). No one truly buys the explanation that it was Falstaff's instinct that prevented him from assaulting his disguised superior, but there is little doubt that everyone is impressed by Falstaff's quickly cobbled-together defense. If it is a skill that can be learned, it is worth Hal's time to have Falstaff teach him. In *2 Henry IV*, Falstaff boasts of his double capacity in the realm of wit: "Men of all sorts take a pride to gird at me. The brain of this foolish-compounded clay, man, is not able to invent anything that tends to laughter more than I invent, or is invented on me. I am not only witty in myself, but the cause that wit is in other men" (I.ii.5–9).

There is little doubt that Falstaff is cause of redirected aggression in men like Hal himself. And these devices of wit will remain with Hal even after Falstaff's departure, though while on stage Falstaff seems forever to hold the upper hand. In *2 Henry IV*, for a second time Hal catches Falstaff behaving badly in public. He confronts Falstaff with, "Yea and you knew me as you did when you ran away by Gad's Hill" (II.iv.279–80), clearly thinking that Falstaff will awkwardly fall back on an old defense. But here, too, Falstaff's wit is fresh and just ahead of Hal, for Falstaff generates a new explanation: "I dispraised him [Hal] before the wicked, that the wicked might not fall in love with him" (II.v.291–93). Here, Falstaff's earlier reported "instinctive fear" of Hal's authority is replaced by his claim that dispraising Hal was a lie designed directly for the benefit of his regal buddy. Again, Hal and we must concede that Falstaff's quick explanation commands a grudging admiration of the cat-like reflexes hidden in this hill of flesh.

But, inevitably, Falstaff's Eastcheap celebration of the flesh is doomed the moment it attempts institutionalization on the larger stage of the English court. So indulgently successful as the changeable rogue of Eastcheap, Falstaff rather ironically fails in the end to adapt to changing times that follow the death of Hal's father. As Hal ascends the throne, Falstaff believes he can cash in, literally, on his investment in the new monarch. That, of course, is expecting too much since Falstaff's misrule cannot find a sanctioned role in the light of legitimate rule. Hal knows this and, surely, Falstaff does too, as he rather sadly begins to acknowledge his debts and bemoan his age toward play's end. When finally officially threatened with banishment by Hal, Falstaff is offered a way to redemption, incorporation, and domestication if he mends his ways. But rather than accept the fact that his amoral, carnivalizing existence must transform to accept the inevitable Lenten turn of events, Falstaff holds to his renegade behavior and hopes that the world will bend to his will rather than the other way around. Falstaff hopes that Hal's stern declarations are merely public pretense and that he will be called for behind the public scene and given his true though unofficial rewards. This does not happen, however. Falstaff never returns to the stage and is, much later in *Henry V*, reported to have died, the result ironically of having become too encrusted in his renegade self-invention. Perhaps just too old to change, he can no longer adapt to new conditions at a point when adaptability really does become necessary for survival.

But Falstaff does in a manner survive the new conditions. We see Hal as beneficiary of the mentalized carnality that Falstaff opened up to him, literally internalizing valuable lessons from his experiences. Recall that prior to Falstaff's banishment, Hal wonders, "Doth it not show vilely in me to desire small beer?" (*2 Henry IV*, II.ii.5–6), alluding to the fact that his tastes now incline toward the grounded commoner rather than toward elevated royalty. Falstaff later observes—and his words are accurate perhaps for the first time—"Prince Harry is valiant; for the cold blood he did naturally inherit of his father he hath, like lean, sterile, and bare land, manured, husbanded, and tilled, with excellent endeavour of drinking good, and good store of fertile sherry, that he is become very hot and valiant" (IV.iii.112–17). Falstaff is describing in Hal a literal humorous transformation, and, correct to a good degree, he claims credit for the transformation. In a typical return to his braggart self, Falstaff resorts immediately to overstatement: "If I had a thousand sons, the first human principle I would teach them should be to forswear thin potations, and to addict themselves to sack" (IV.iii.117–19). Truth inheres in Falstaff's message, however. Hal's biological father argues that Hal's presence among commoners has cheapened Hal as heir to the throne.

Clearly, Hal stands in contrast to his brother, John, who is more than capable of taking the same Machiavellian political road of deception that their father took on his way to the throne by gulling opponents into ignoble death traps. Clearly, Hal is not of the same order as John, likely because of his escapades in Falstaff's Eastcheap, and that is a good thing. Falstaff's opinion of John hits home: "Good faith, this same young sober-blooded boy doth not love me, nor a man cannot make him laugh. But that's no marvel; he drinks no wine. There's never none of these demure boys come to any proof; for thin drink doth so overcool their blood, and making many fish meals, they fall into a kind of male green-sickness" (IV.iii.84–90). According to Falstaff, John lacks physical vitality, while Hal's own vitality has been nurtured by exposure to the unrepentant, corporeally indulgent Falstaff. One could accurately modify Falstaff's "banishment" proclamation to make this even more accurate point: Love not Falstaff and love not the world. Failing to indulge the world leads to an enfeeblement that Falstaff speaks of concerning John and also that the post–*Richard II* plays reveal of Hal's father, Henry IV. Henry decayed into a man both physically and spiritually sick as his actions weakened and sickened his country under him.

But Hal is more than both of these blood relations, and it's arguably because of his humorous romps in Eastcheap, his environmental exposure to the "low" life that has humorally adjusted what appears to be a familial predisposition to phlegmatic distance from life and living. Knowing his citizenry beyond the class-conscious confines required of royal behavior, bowing to misrule as he has, Hal comes to know those he will command in battle and rule from the throne. Climbing down from the high altar of idealized royalty and experiencing the material and "suspect" body of his subjects, Hal comes to know their needs and motives. And he uses that special awareness during his brief but glorious reign of England in ways his predecessors and successors did not and could not. Failing to know their subjects in this manner, these other kings' reigns are troubled in ways his never is.

Falstaff provides an overwhelming adjustment to Hal's apparent genetic inheritance, without which would occur a dreary continuation of the same sort of rulership that had impaired England for decades. Most impressively, Hal learns from Falstaff something Falstaff himself sadly forgets at the moment of greatest need: that perpetual adaptation is a first rule of survival. Finally, encrusted in his own manufactured persona as unwavering lord of misrule, Falstaff fades from the scene. Adaptable to the extreme, Hal converts to Henry when the time is ripe. Was he pretending to be other than himself all along while in Eastcheap? It's really an irrelevant, Hamlet-like question since the difference between being a nobleman pretending to be a drunken brigand and being a brigand pretending to be noble are of little real difference (or profit) to the world.

If Falstaff lives on the larger stage of English life in Hal, he lives in yet another manner as well. It is regularly surmised that the personified Epilogue in *2 Henry IV* was performed in Shakespeare's day by the actor who played Falstaff, itself something of a leasing resurrection of our old friend. Epilogue observes of this "unpleasing play" that in the next installment "our humble author will continue the story with Sir John in it, and make you merry with fair Catherine of France" (*Epilogue*, 23–25), apparently trying to allay audience concerns that Falstaff is no more. John Falstaff, however, doesn't return in the subsequent *Henry V*, and Catherine's appearance is not the promised merry one. And that's probably just as well. Epilogue here unrealistically reflects a wish for an unending season of indulgence, which cannot be in this changing world. The business of the world must move forward. But in this case, the shadow of Falstaff's flesh will loom over that world in a way that even Epilogue fails to foresee. Though literally the story ends poorly for Falstaff, his triumph lingers well after the events in *Henry V* fall victim to audience forgetfulness. Well after the plot details of these Henry plays proper are forgotten, the memory of Falstaff's vitality remains available to serve the audience as much as it does Hal upon his transformation to King Henry V. As Wylie Sypher notes, "Falstaff proves what Freud suspected: that comedy is a process of safeguarding pleasure against the denials of reason, which is wary of pleasure. Man cannot live by reason alone or forever under the rod of moral obligation, the admonition of the superego. . . . Comedy is a momentary and publicly useful resistance to authority and an escape from its pressures."[15] Public utility reveals comedy's intrinsic value. We are *reminded* of our place by testing the limits that place delineates through the resistant devices of comic embodiment. We must remember, too, that an unadulterated form comedy can only have a limited life. In fact, its "momentary" nature gives comedy its intensity, a point Hal himself brings home when he muses, "If all the year were playing holidays, / To sport would be as tedious as work" (*I Henry IV*, I.ii.157–58).

Bakhtin and Rabelais' World

Falstaff's main problem is that his Eastcheap carnival existence had a limited life expectancy, an unavoidable point that Falstaff never willingly faces. As Erich Segal reports, "Like Odysseus after his ten-year orgy of outrageous behavior, however freely the celebrants may behave during the festival, when it is over they return to the order of the everyday world."[16] Exactly how long Falstaff's misrule held sway over Eastcheap is unclear. It may have extended even longer than Odysseus' ten-year orgy. But inevitably it did come to an end, following an inevitable pattern we see formalized by Christian pre-Lenten festivities rounded by the "civilized" social and political order of the day. The Mardi Gras climax of the carnival season that begins with Twelfth Night—celebrating the gift-giving of the Magi—always ends with somber Ash Wednesday. Even Shakespeare's celebrated *Twelfth Night*, set in a world whose business is anything but business, ends with the clown's somber song, returning us to the humdrum world of dreary, repetitive drudgery: "For the rain it raineth every day" (V.i.138).

But significantly, despite the inevitable return to the somber/sober light of day, carnival festivities—Falstaff's as well as others—have the full potential to enact quite significant change, though their contributions to the sanity of the world may not always be fully evident. This is exactly what we see when Hal absorbs Falstaff's Eastcheap misrule into himself and brings its salutary effects into the workaday existence of English politics and ruling-class society. Comic celebration and carnival may be by definition unstable, but they go far to counter stultifying effects of static, stratified, orderly, ideal, and idealized institutions.

Mikhail Bakhtin's articulation of the medieval carnivalesque vision offers crucial insights into Falstaff and the valuable effect of his comic like by offering a contextualization of this brand of misrule in contrast to rising dualist orthodoxy found in the Renaissance and later. Bakhtin centrally observes that "[l]aughter purifies the consciousness of men from false seriousness, from dogmatism, from all confusing emotions."[17] Studying in particular the works of François Rabelais (ca. 1490–1553), Bakhtin expands on the point by quoting from L. E. Pinsky: "For Rabelais, man of the Renaissance, laughter was precisely a liberation of emotions that dim the knowledge of life. Laughter proves the existence of clear spiritual vision and bestows it. Awareness of the comic and reason are the two attributes of human nature. Truth reveals itself with a smile when man abides in a nonanxious, joyful, comic mood."[18] Laughter liberates humanity from the shackles it has placed on itself. It is a tool generated by human nature, and when unfettered to do its "job," as Susanne Langer observes, "real comedy sets up in the audience a sense of general exhilaration, because it presents the very image of 'livingness.'"[19] Unadulterated by self-conscious control, laughter brings us to a level of liberated "livingness." To this point, Nathan Scott reminds us of a crucial related observation about comedy: "It is not, of course, the purpose of the comedian to enforce a simple Sunday-School lesson: all he wants to do is to give his suffrage to the Whole Truth."[20] Improper though it must at times be, comedy does not fundamentally serve a moralizing purpose. If it is important to accept that comedy offers a view of a level of "livingness" we sometimes overlook, it is crucial to recall also that moralizing—and sometimes even ethics/morality itself—frequently (if not invariably) stands stultifyingly in the way of the comic mission.

Even the "proper" Victorian George Meredith touches on the potential amoralistic subversiveness of comedy. For the most part, he rather stodgily observes that the "laughter of comedy is impersonal and of unrivalled politeness, nearer a smile—often no more than a smile."[21] But he also rather curiously pleads for a return of the kind of broad humor and comedy found in Aristophanes, Rabelais, Voltaire, Cervantes, Fielding, and Molière. And from that point he even identifies the "target" of comedy in Rabelaisian, rather than moralistic, terms: "We have in this world men whom Rabelais would call agelasts; that is to say, non-laughers; men who are in that respect as dead bodies, which if you prick them do not bleed. The old grey boulderstone that has finished its peregrination from the rock to the valley, is as easily to be set rolling up again as these men laughing. It is but one step from being an agelastic to misogelastic, and the . . . the laughter hat[er] soon learns to dignify his dislike as an objection in morality" (114–15). Unable to feel life coursing through his veins and lacking any human elasticity, the agelast is culture's blocking character, the creature that insists on stability at all cost.

Though the implication is that agelasts are crusty old men, Troilus and the misanthropic Alceste are evidence that superannuation is not a necessary quality. In fact, Shakespeare's *Measure for Measure*, dominated by a youthful but coldly moralistic Ans"elo, perfectly exemplifies the danger of having an agelast in our midst, not to mention an agelast possessing the youthful vigor and given the authority to enforce agelastic restrictions. As Angelo's master, the Duke of Vienna observes,

> Lord Angelo is precise,
> Stands at a guard with envy; scarce confesses
> That his blood flows, or that his appetite
> Is more to bread than stone. Hence shall we see,
> If power change purpose, what our seemers be. (I.iii.50–54)

Angelo's initial sin is what comedy identifies as the central sin of personal vanity, out of which arises, yet again, the matter of seeming and being as central concerns. Angelo seems virtuous, but it is a posture he can't sustain, despite his overweening pride, as his moralistic predisposition compels him to succumb to dastardly dissembling in order to satisfy his "unseemly" bodily desires.

In the case of *Measure for Measure*, however, the unmasking of the agelast does not involve laughter to open the way to the truth. This play is too mired in near-agelast victory to let in too much humored light. What were likely to be once carnivalesque times in *Measure for Measure*'s Vienna have run too long and have decayed by overindulgence, turning to joyless licentiousness and sex for hire. But silencing the body is never the proper answer. What *Measure for Measure* reveals is the persistent call of the body to resist full domestication, sometimes even without the vital assistance of laughter. So instead of threats and rapacious coercion to satisfy his lust under cover of night, why didn't Angelo "go natural" and simply express his affections for Isabella? Surely, what seems to be a growing affection in Angelo for Isabella is not a bad thing. Wouldn't taking a traditional approach to love have been a chancy though far more acceptable alternative to the actions he chooses to take? Caught in his own web of self-importance, Angelo can only turn to hypocrisy and maintain his encrusted façade of being uninfluenced by the "lower" urges that have overwhelmed his Viennese subjects. He must hold to his inflexibly rigid social self while pursuing the bodily urges that clearly clash with the abstracted ideals he strives to personify and enforce in Vienna. Failed or refused negotiations with the body will unmask the dissembler and reveal agelasm in the fullness of its deadening hypocrisy.

Malvolio in *Twelfth Night* falls victim to similar arrogance. But unlike the bleak and glum world of *Measure for Measure*, the world Malvolio occupies falls colorfully under the influence of laughter-generating carnival, and this agelast is exposed via sport and game-ish foolery, forced to concede his failings as everyone else safely has a good laugh. Though in weight and literal substance not as big as Falstaff, Sir Toby Belch effectively misrules over events in direct confrontation with agelastic seriousness. And while Malvolio does not change from agelastic moralizer to engaged carnivalizer, the aura of agelasm departs as Malvolio departs from Illyria and the various lovers are united. Drawing toward the close of the play, following revelations of Malvolio's mistreatment, Fabian observes,

> How with a sportful malice it was followed
> May rather pluck on laughter than revenge,
> If that the injuries be justly weighed
> That have on both sides passed. (V.i.355–58)

Despite suggestions that everyone is willing to forgive and forget, Malvolio chooses not the release of forgiving laughter, but opts to pursue satisfaction in the form of a deferred revenge. Agelasts themselves are rarely won over, and as a result the carnival clown will never go out of business.

Whether or not it triumphs, however, laughter is the best defense against the agelast and his encrusting, moralistic influence over worldly affairs. As

Meredith observes, the comic mission is somehow to breathe life into the virtually dead, either by resurrecting bloodless, humorless creatures themselves or by revitalizing the stultifying, bloodless world that their pervasive humorlessness has molded. Here is where Rabelais' example of *grotesque realism* is of crucial importance. It is a form that, according to Bakhtin, encourages an open-endedness that is at the heart of carnival: "Rabelais' images have a certain undestroyable nonofficial nature. No dogma, no authoritarianism, no narrow-minded seriousness can coexist with Rabelaisian images; these images are opposed to all that is finished and polished, to all pomposity, to every ready-made solution in the sphere of thought and world outlook" (3). Closure, finite systems, order, and controlling authority of all sorts are the objects of Rabelaisian assault.

The universe of Rabelais' grotesque realism is an expanding, infinite universe rather than the static creation imagined by hierarchic orthodoxy. And the open-ended, dynamic, physical *body* stands forth as evidence of this vision, in stark contrast to the static vision of idealized being that was expostulated by Renaissance authority (and articulated so punctiliously in *Troilus and Cressida*). In the authorized Renaissance mind, Bakhtin observes that "the body was first of all a strictly completed, finished product." He continues, "All signs of its unfinished character, of its growth and proliferation were eliminated; its protuberances and offshoots were removed, its convexities (signs of new sprouts and buds) smoothed out, its apertures closed" (29). This may have been official doctrine during the Renaissance, but it is manifest even today in the obsession we currently have with the tidying up processes of cosmetic surgery.

Grotesque realism, on the other hand, provided a popular, subterranean alternative to dominant culture's encrusting tyranny over the open body: "In grotesque realism . . . the bodily element is deeply positive. It is presented not in a private, egotistic form, severed from other spheres of life, but as something universal, representing all people. As such it is opposed to severance from the material and bodily roots of the world; it makes no pretense to renunciation of the earthly, or independence of the earth and the body" (Bakhtin, 19). The body is a universal connector to our environment and also to the rest of humanity, given that it is clearly something that all humanity has in common. As such, grotesque realism takes on a task larger than merely aggrandizing the autonomous body. In the process of aggrandizement, the body actually honors the points of commonality inhering in all humanity, inviting a recognition of our interrelationships with our full environment. This anti-autonomous sense of body, in turn, sinks to a level that celebrates the "degraded" components of the body, "the lower stratum of the body, the life of the belly and the reproductive organs" (Bakhtin, 21).

But it is degradation not in the sense of "high" and "low" so much as it is the sense of bringing value, quite literally, down to earth: "To degrade an object does not imply merely hurling it into the void of nonexistence, into absolute destruction, but to hurl it down to the reproductive lower stratum, the zone in which conception and a new birth takes place. Grotesque realism knows no other level; it is the fruitful earth and the womb. It is always conceiving" (Bakhtin, 21). Escaping from the tyranny of orthodox restrictive idealism, steeping itself into its own basic regenerative functions, the body provides a groundwork on which new orders, mental and otherwise, may arise.

Grotesque realism works to defeat autonomous agelasm, which is little more than a variant description of idealistic, Platonic subjectivity, the putative cornerstone of Western culture—philosophy, theology, and psychoanalysis certainly included. Humbling our egoistic wills/ideals by indulging our appetitiveness, grotesque realism works to reveal an understanding (and acceptance) of our common material heritage first by noting two traditions manifest most clearly in the European Renaissance: "[O]ne of them ascends to the folks culture of humour, while the other is the bourgeois conception of the completed atomized being. The conflict of these two contradictory trends in the interpretation of the bodily principle is typical of Renaissance realism. The ever-growing, inexhaustible, ever-laughing principle which uncrowns and renews is combined with its opposite: the petty, inert 'material principle' of class society" (Bakhtin, 24). The folk tradition celebrates in particular bodily apertures and orifices most open to the rest of the world, manifest most memorably and "lowly" in the expulsion of body sounds, especially at somber, "closed," and therefore inopportune moments. We laugh because the body violates and exposes our delusionally abstracted sense of "ideal" autonomous selfhood, which dresses up and tries to hide our biological selves. Abuses, curses, and oaths are echoes of this celebration as well. But laughter seems the master.

Bakhtin's celebration of Rabelais' celebration of the grotesque and carnivalesque led Bakhtin to conclude that "[t]he men of the Middle Ages participated in two lives: the official and the carnival life" (96). And he observes, "Carnival (and we repeat that we use this word in its broadest sense) did liberate human consciousness and permit a new outlook . . . ; it had a positive character because it disclosed the abundant material principle, change and becoming, the irresistible triumph of the new immortal people" (274). Most historians now see the official and carnival lives distinction as overstated, given that the carnival life was itself recognized and even sanctioned by "official culture." Bakhtin's dichotomizing vision of medieval life led him further to argue that, through the carnivalesque, humanity had a brief taste of the ideal human condition—a new immortality—toward which we even today should be striving. It is a logical enough conclusion to draw, but it seems as overstated as his polarized world view in general, for it dreams of a utopian, postlapsarian return to a prelapsarian state of nature that evolutionary science has frankly demonstrated never existed. The future may evolve in unexpected ways, but banking on dreams and visions is a shaky business. Comedy does as much at times, but it would also do well to propose more viable alternatives even as it dreams of utopia.

The Orifice of the Body Politic: Society's Marketplace

As he discussed the concept of the egoless body's immortal potential and "all-people's character" (19), Bakhtin also reminded us of the central importance of the marketplace to populist, carnivalesque challenges to dominant culture, observing that "all the great writing of the Renaissance [captured] the carnival atmosphere, the free winds blowing from the marketplace" (275). Bakhtin's celebration of the open body and all its bodily functions seems almost naturally to lead to the marketplace. Given that celebrations generally attend market days, and given that markets serve the body in more direct ways than any other social institution, the connection seems almost unavoidable. This place where the body seeks food for survival also becomes the open-orificed gathering place for irreverent behavior that implicitly challenges official restrictions on such activities as labor, productivity, and distribution.

The marketplace is simultaneously a site of centralized social control as well as a site of socially ordered vulnerability. Here, the bounty of nature is converted by social forces to market value and incorporated/absorbed into the social fabric. But in the process of converting raw material to commodities with assignable value, a certain uncontrollable power is conferred upon the agrarian custodians of bountiful nature in the bargain. The market is a point of potentially dangerous/uncontrolled exchange between orthodox culture and "othered" nature, wherein culture is forced to acknowledge the value of nature's bounty at that vulnerable moment of negotiation when it attempts to absorb/digest that bounty into its system of order. Prior to the market, nature's bounty stands outside the power structure; and after the market, its concentrated power is officially domesticated, dispersed, and consumed. But at this moment of marketplace concentration, nature can use its tantalizing bounty momentarily to hold hostage the ordered social system that it would subsequently nourish and sustain. This is the moment that the culture of folk humor holds the reigns of civilization itself. The fertile abundance concentrated in the marketplace generates an atmosphere that is potentially regenerative and positive on multiple levels.

With a sort of intellectual rigor and cultivated wit that is Ben Jonson's trademark, his play *The Alchemist* (1610) reveals itself to be part of this iconoclastic folk tradition, though he adds a unique twist to the tradition of the marketplace. Never satisfied to follow even iconoclastic tradition, Jonson takes a slightly altered route and adds a notable wrinkle to the tradition he taps into. Rather than beginning from the position that opens onto marketplace bounty, he begins with a world that is physically sick, literally and metaphorically infected with the open buboes of plague and sickness. The cure, Jonson seems to argue, will come from carnivalesque forces found in the market, from exchange and redistribution of wealth that are the cornerstone of market enterprise.

In a parody of the celebratory opportunities provided by carnival season, Jonson's "season" of plague provides like opportunity. The plague invades the body politic as it does the actual human body, opening up and disrupting every closed and orderly system it contacts. At the social level, its disrupting influence cataclysmically forces a cultural adjustment to a system grown vulnerably stagnant, isolated, and arrogantly unresponsive to its environment. This is the same vision embraced by Antonin Artaud, who asserted that plague "releases conflicts, disengages powers, liberates possibilities, and if these possibilities and these powers are dark, it is the fault not of the plague nor of theatre, but of life."[22] Inserting such dark sentiments into comedy, Jonson's *The Alchemist* engages carnival disruption by way of the dark parallel presence of plague. Though not unlike Shakespeare's *Measure for Measure* in many regards, the result here is that in *The Alchemist*, the light shining through the darkness creates a memorably comic cultural dawn.

Jonson's London in *The Alchemist* is on the waning side of a plague outbreak, but the physical, social effects of the outbreak are still in place: lords and masters have left the city to the *mis*rule of the city's "lesser" denizens, an abandonment that has allowed for the infestation of the city by myriad social parasites. The opportunity to operate with a liberty unchecked by codes and laws is exactly what truly free enterprise and industry seek. In this case, plague opens the city to unchecked, raucous market activity among those brave enough or foolish enough to remain behind. However, rather than stepping into a marketplace that converts nature's bounty into value exchange, the market introduced in *The Alchemist* is an empty grotesquerie of an actual market. The products on sale have no real value whatsoever, being merely promises of alchemical windfalls ultimately designed to gull the fanciful and avaricious. In this regard, *The Alchemist* mischievously undermines even the mischievous marketplace, uncovering a doubly dense bounty of chaos feeding on chaos with order nowhere in sight.

Face, the servant of the absent master Lovewit, has taken the opportunity of his freedom from his master to recruit two "business" acquaintances, Subtle and Dol Common. They will engage in a parody of a traditional market exercise by producing products of "value" virtually out of thin air, seeking to generate market demand motivated by the humorous imbalance of avarice. The alchemical enterprise they pretend to master involves converting naturally base and imperfect matter to products of *ideal* perfection that emulate, but only by parody, labor-intensive processes that lead to natural bounty. Since money doesn't grow on trees in the natural world, the mission of alchemy is to provide a process that replaces the fruited tree by directly transforming base material to ideal value without the tedious, labor-intensive task of cultivating the land or husbanding animals. Converting base metals to gold is just one harvest they claim to undertake. Other transformations include harnessing luck and chance in the arenas of gaming, love, and matrimony, and converting louts to gentlemen. The carnivalesque arises as these three merchants of alchemical pretense transform the airy nothings they claim to possess into the capital gains that will secure their futures at the expense of those gulls they pretend to help. This in the fullest sense is their alchemical magic. While on the surface they pretend to provide their customers with undreamt of alchemically generated wealth, power, and success, it is their own "alchemical" accumulation of wealth, power, and success that we witness.

Metaphorically speaking, this too was Falstaff's goal, to gather a living he did not materially earn by fruitful labor, choosing instead to utilize his wit and native ingenuity *alchemically* to accumulate material products of value and creature comfort from those who were in possession of such comfort but did not (according to Falstaff) deserve it—or at least deserved it less than he did. If a fool and his money are soon parted, Falstaff is happily available to catalyze this transfer. Face, Subtle, and Dol Common are like Falstaff but are far more deliberate, if not more successful. Face is in many ways more a master of the individuated psychology of greed than Falstaff could ever hope to be. Like a seamy savvy door-to-door salesman, he walks through London and reads in the citizenry instances of unsatisfied personal desire/hunger, winning over whomever he meets with the promise that he can make dreams come true even to the point of unimagined overabundance. And while there is not a character among the list of customers whose desires are not in one way or another "noble," it is their extreme lust for the unearned personal fortune of quick and easy profit that blinds them to the ruses of Face and his cohorts. Longing for more than they deserve, the customers ultimately get exactly what they deserve, namely to be gulled out of the very materials they put up as collateral for the lottery-like profit they hope to win. Each provides a valuable lesson befitting the best of moralistically satiric writers: aspiring for unearned advancements beyond one's actual capacities guarantees a fall rather than an elevation.

However, Jonson does more in this play than merely create an effective and satisfying satire on avarice. Just as the gulling approaches its alchemical perfection—just as the three flimflam artists are about to escape with what a twisted morality could call their justly earned wages—Face's master, Lovewit, unexpectedly returns from the country. At this carnivalesque peak, Lovewit discovers Face's ingenious duplicity. But rather than punish Face, Lovewit enters the game, backs Face against all charges levied by the gulls, and turns away all the victims, humbled and forced to digest the folly of their ways, a good deal lighter in their pockets in the bargain. The young Lovewit gains a wife (the widowed Dame Pliant) and a store of valuable properties.

Even more than that, Lovewit gains a new relationship with his servant, Face, to whom he says near closing (calling him for the first time by his first name), "I will be ruled by thee in anything, Jeremy" (V.iii). Lovewit concludes,

> That master
> That had received such happiness by a servant,
> In such a widow, and with so much wealth,
> Were very ungrateful, if he would not be
> A little indulgent to that servant's wit,
> And help his fortune, though with some small strain
> Of his own candor. (V.iii)

Telling Face, "Speak for thyself, knave," Lovewit yields the stage to Face, who confesses to the audience in closing:

> My part a little fell in this last scene,
> Yet 'twas decorum. And though I am clean
> Got off from . . . all
> With whom I traded; yet I put myself
> On you, that are my country: and this pelf,
> Which I have got, if you do quit me, rests
> To feast you often, and invite new guests. (V.iii)

The "pelf," or ill-gotten gains, is served up to the audience, whose profit has been not merely a tale with a rather dubious and uncertain moral, but a glimpse of ingenious misrule and disorder rewarded in the returning order with promotion, new-found security, and respect.

In *The Alchemist*, order has been disrupted by plague and ensuing market-place misrule. Jeremy Face's value as an ingenious man of depth—more than a mere face—is clearly established. When order to London finally returns, this master-servant relationship will be of a new order altogether, involving mutual respect and indebtedness. Subtle and Dol Common, trusting to a dubious doctrine of honor among thieves, have been thwarted without benefit of integration into the new order. Evicted without the fruits of their duplicitous labor, at least they depart with their skins. But Jonson allows carnival to play itself to climax and rewards the lord of misrule with an honored place in the renewed order, while those victimized by their own unbalanced humors are left to ponder events, Subtle and Dol included. The plague ends, flesh wounds heal, and a new health spreads through a newly awakened London.

Following in Jonson's distant footsteps and taking the comic reliance on marketplace liberties to a still different level, John Gay's *The Beggar's Opera* (1728) stands out as the folk-tradition singularity, tapping into much of what Bakhtin identifies as crucial to the Rabelaisian Renaissance tradition, especially the festive tradition of grotesque realism. The world Gay creates rivals Jonson's London and Shakespeare's Eastcheap, full of unrepentant sinners lusting for life outside the social order of the day. But with Gay, celebrating the low life takes on an immediately darker hue, much more in line with the Vienna of Shakespeare's Angelo. Gay's festive vitality attaches to an underworld of cutpurses, highwaymen, fences, and prostitutes that decidedly opposes official culture by indulging unofficial *mis*rule. But even in this world generally beyond the reach of official culture, official culture still casts a long shadow. What we see is festive vitality consistently jeopardized by agelasm, making a case for its pervasiveness in the human consciousness and even its sinister self-preserving allure. With Gay in particular, we see that the cards are stacked against the comic spirit from the start.

Even Gay's roguishly indulgent hero Macheath is haunted by dark visions, gloomily conscious of the transitory nature of indulgent pleasure despite the fact that his "dream" of a harem is actually realized:

> Youth's the season made for joys,
> Love then is our duty;
> She alone who that employs,
> Well deserves her beauty.
> Let's be gay
> While we may,
> Beauty's a flower, despised in decay.
>
> Let us drink and sport today,
> Ours is not tomorrow.

> Love with youth flies swift away,
> Age is nought but sorrow.
> Dance and sing,
> Time's on the wing,
> Life never knows the return of spring. (II.iv.Air 22)[23]

Macheath has an official culture's concern that youth ultimately yields to decay, stagnation, and death.

In *The Beggar's Opera*, darker concerns of mortality and self-interested acts of greed perpetually hover over its celebrants of misrule. For example, Macheath is plagued throughout the play by the specter of bigamy, or maybe even polygamy, infiltrations of social sanction that destroy his ease. If Macheath can in his libertine way claim, "I must have women. There's nothing unbends the mind like them" (II.iv.21–22), he is haunted by an awareness of the restrictive cultural sanctions of matrimony, which in the play are revealed to be of economic import far more than of romantic significance. The underworld mastermind Peachum makes clear the point that love has little real relevance in a world governed by the ideals of ownership and control that infect even this world of literal *out*laws. When his daughter Polly confesses of Macheath, "I love him, sir; how then could I have thoughts of parting with him?" (I.xi.21–22), Peachum reveals his understanding of the notion of love: "Parting with him! Why, that is the whole scheme and intention of all marriage articles. The comfortable estate of widowhood is the only hope that keeps up a wife's spirits. Where is the woman who would scruple to be a wife, if she had it in her power to be a widow whenever she pleased? If you have any views of this sort, Polly, I shall think the match not so very unreasonable" (I.xi.23–30). Profit through widowhood is the only worthy part of matrimony. When it is decided that Macheath must die, Peachum almost waxes sentimental: "[I]t grieves one's heart to take off a great man. When I consider his personal bravery, his fine stratagem, how much we have already got by him, and how much more we may get, methinks I can't find in my heart to have a hand in his death" (I.xi.6–11). Friendship itself hinges on matters of profit. And since, as Peachum notes, "Macheath may hang his father and mother-in-law, in hope to get into their daughter's fortune" (I.viii.65–67), one criminal must peach first on the other in order to secure his future.

In short, in *The Beggar's Opera*, the abundant market of folk tradition has been overtaken by the restrictive, dominant-culture, capitalist market of commodity exchange. In *Beggar's Opera*, Gay peels away the trappings of polite society by going underground and shows us by indirection that market culture is governed by greed and avarice rather than by equity in exchange, that a successful (i.e., profitable) transaction is little more than sanctioned theft. Love itself is a commodity, and friendship little more than a tenuous business partnership. This seamless application of capitalist dogma to what should be scruples-free underworld enterprises is an ingenious inversion of orthodoxy that exposes the orthodoxy for what it is, much as the heroism of the Trojan Wars could be exposed for being little more than a ten-year exercise in greed-motivated plunder.

The Beggar's Opera, as many have noted, is in large part a stinging satire of eighteenth-century London, replacing the scaffolding of polite society with a stripped-down mirror image of that society aped by an undersociety coordinated by a scheming lord of misrule who himself lives in close proximity to the gallows, put to use not only by the victimized polite society but by members of the victimizing underworld as well. Peaching to the authorities on the underworld competition even seems a good strategy for underworld advancement, for there's always the hope that when fortune shines for an underworld boss, "[a] rich rogue now-a-days is fit company for any gentleman" (I.ix.10–11). The fruits of this brutal underworld market exchange can actually effect a leap to respectability. Finally, we see that the codes and general structure that support polite society are revealed to be devoid of honorable intentions and therefore no different than mere criminality. Their applications alone to human social units guarantee little more than a world of mechanical relationships able to be located even in the lowest stratum of human society, the stratum that should be least influenced by the pull of orthodoxy. Surely, there must be more to human existence than what is here so baldly presented.

That "more" could very well be the love that Polly and Lucy are capable of feeling. They appear to be fundamentally different from their parents, even different from the other women in the play. Sadly for them, however, their love is given to the same man, who is finally "unworthy" of either of their loves. But this is a moralizing conclusion. One could ask why these women are so foolish as to choose Macheath rather than, say, a respectable "gentleman" perhaps more to their upwardly ambitious parents' liking. This, of course, brings us to the kinds of plots abundant in eighteenth-century novels and plays, wherein invariably a virtuous *and* acceptable mate is finally located. Often we discover that the illegitimate rogue has legitimate roots or that he undergoes a moral transformation just before all is lost. But this is not Gay's intent. Rather, he refuses to solve Macheath's bigamy/polygamy dilemma and allows Macheath to be condemned to die. Even a sub-society divested of its refined trappings cannot perpetually accept the indulgent disorder of unregulated carnival indulgence, it seems.

Here's where Gay's play does its own spark of carnivalesque alchemy and infects the social fabric despite its perhaps too heavy moralizing embroilments. Without formally showing how to bring it about, Gay "argues" that social order must bend but not break in its confrontation with festive indulgence derived from true love disentangled from its commercial dimension. At least society must allow for release of the tensions that arise from the conflicting manifestations of love as festively open and socially sanctioned/controlled. Through song, abundant feasting, brazen defiance of order, and general "indecency" throughout *The Beggar's Opera*, the carnival of the open marketplace vitalizes into the playhouse proper by opening receptive "orifices" through laughter.

The Beggar's Opera in the end is "fun" even as it seems to have painted itself into a moralistic corner by having Macheath "justly" condemned to death for his myriad crimes against society. And in the service of that fun-endorsing, amoralistically anti-agelastic open-ended end, Gay offers a rather unlikely ending to the play. Before Macheath actually dies on the scaffold for his transgressions of orthodox codes, Player and Beggar (whose opera it is, after all) interrupt the performance onstage and rewrite it before the audience's eyes. Beggar confirms the stern original intent of the play: "Had the play remained as I at first intended, it would have carried a most excellent moral. 'Twould have shown that the lower sort of people have their vices in a degree as well as the rich; and that they are punished for them" (III.xvii.24–28). Beggar and Player have allowed the play's severe moral almost to run its course but now choose another path to follow. Player complains that the unhappy moral ending—hanging Macheath—is unacceptable. And Beggar agrees: "Your objection, sir, is very just and easily removed. For you must allow that in this kind of drama 'tis no matter how absurdly things are brought about—So—you rabble there—run and cry reprieve!—let the prisoner be brought back to his wives in triumph" (III.xvii.11–16). This new ending, complete with reprieve, "solves" nothing, of course, in the world of just, retributive order. Everything remains the same: human greed has not been vanquished; human folly has not found a corrective alternative path. But could it be that vitality itself should be considered a just defense against agelastic punishment?

Macheath closes with, "[T]hink of this maxim, and put off your sorrow, / The wretch of today may be happy tomorrow" (III.xvii.19–20). The play, in the end, reveals that such a turn of events literally requires an illogical breach of socially sanctioned impositions of order. Things *may* change for the better but only through an interruption of the processes of social interaction as currently constructed. It is always possible to wish that some sort of social transformation will occur, endorsing, say, a gift of forgiveness (grace or mercy) to disrupt agelastic severity. Along these lines, Sypher identifies a parallel function found in forms generally affiliated with the comic vision: "The joke and the dream incongruously distort the logic of our rational life. The joke and the dream are 'interruptions' in the pattern of our consciousness. So also, possibly, is any truly creative work of art a form of 'interruption' of our normal patterns or designs of seeing and speaking, which are mere formulas written on the surface layer of the mind."[24] Jokes, dreams, art: to this list surely should be added the comic lawlessness of Rabelaisian carnival, in its varied manifestations on the stage. Song, dance, and flying beneath the radar of law and order, exuberance leaves its trace upon us even as "right behavior" moralizes itself back into primary position. The matter now at hand is how to use that "trace" for more lingering transformative purposes.

CHAPTER 4

More than Matter Matters

When, in 1978, Robert M. Torrance identified the distinction between the two basic comic strains, he also observed that "the two conceptions of comedy, as satire and as celebration, though opposite, are by no means exclusive. . . . Satire and celebration can be, and normally are, interwoven, and no small part of the comic hero's valor is revealed in his spirited capacity to surmount the derision that continually befalls him."[1] While *The Beggar's Opera* does "interweave" these two components of comedy, it also demonstrates something of the vinegar-and-wine quality that so many such works possess. Generally, the comic hero remains outside the perhaps chastened but still unchanged social fabric, enjoys a sort of victory even in defeat, or is artificially incorporated into the fabric through some element of artifice. It's hardly a universally beneficial interweaving that occurs.

With *The Threepenny Opera* (1928), a reworking of John Gay's 1728 *The Beggar's Opera*, Berthold Brecht works to interweave the two strains more seamlessly, manipulating Gay's satiric assault on capitalist society and then directly arguing that living society functions best only if and when the body itself is served first. Brecht the materialist Marxist derives as much from folkloric traditions rooted in the Rabelaisian movement Bakhtin identifies as he does from Marxist polemics. And issues of materialist transformation meet polemical affirmation in a mix that reveals both strengths and weaknesses inherent in Rabelaisian celebratory comedy as well as in the Marxist philosophy that Brecht draws out of this comic strain. These ostensible weaknesses include the sorts of concerns that Simon Critchley (and others) confronts when he observes that Bakhtin's vision can errantly "lead to a romanticization and heroization of the body"[2] if one fails to see the need for countervailing influences. And, by extension, an inherent flaw in idealized Marxism surfaces as well.

"Man's Not Good Enough Just Yet"

In *The Threepenny Opera*, Low-Dive Jenny's "Solomon Song" speaks of noble and ideal qualities found in four historical figures (Brecht himself being the fourth), qualities that bring each personage to despair or ruin: "It's wisdom that brought him to this state"[3] in the case of Solomon, beauty for Cleopatra, courage for Caesar, and inquisitiveness for Brecht. Each stanza ends, commenting on the idea of noble aspirations, "How fortunate the man with none" (126). Rather ironically, Macheath is the subject of the fifth stanza of this song, and his virtue is "sexual urges" (127). While indulging such urges appears to many of us as anything but noble, this opposite-end "virtue" receives the same final thought: "How fortunate the man with none." And according to Jenny at least, all of these "virtues" do little more than place individual human survival in jeopardy.

Macheath's "Ballade of Good Living" makes a similar argument, this time unabashedly endorsing material gain, its refrain being, "One must live well to know what living is" (109). In the song, Macheath confesses an inability to be a high-minded dreamer and poet and an unwillingness to suffer in order to achieve nobleness:

> I've heard them praising single-minded spirits
> Whose empty stomachs show they live for knowledge
> In rat-infested shacks awash with ullage.
> I'm all for culture, but there are some limits.
> The simple life is fine for those it suits.
> I don't find, for my part, that it attracts. (109)

And in the third stanza: "Suffering ennobles, but it can depress. / The paths of glory lead but to the grave" (109). Peachum's "Song of the Insufficiency of Human Endeavor" observes the futility of the life of the mind:

> Mankind lives by its head
> Its head won't see it through
> Inspect your own. What lives off that?
> At most a louse or two.
>
> For this bleak existence
> Man is never sharp enough.
> Hence his weak resistance
> To its tricks and bluff. (124)

The "First Threepenny Finale" (ending act I) includes the following lines:

> Let's practice goodness: who would disagree?
> But sadly on this planet while we're waiting
> The means are meager and the morals low. (95)

And later, "We should aim high instead of low. / But our condition's such this can't be so" (96). In the "Second Threepenny Finale" (ending act II), Macheath and Jenny sing:

> However much you twist, whatever lies you tell
> Food is the first thing. Morals follow on. . . .
>
> What keeps man alive? The fact that millions
> Are daily tortured, stifled, punished, silenced, oppressed.
> Mankind keeps alive thanks to its brilliance
> In keeping humanity repressed. (117)

The urge for a noble life invariably confronts the need first merely to survive, confronted in turn by a presumed need to dominate and subdue competing others for self-survival to occur.

While Gay never quite gets to the point of a higher end—backing off from making any firm, direct point—Brecht uses his play's carnivalesque to articulate various actual agenda-driven discourses. For Brecht, his oft-stated higher end is to argue the obstructionism of current capitalist market distribution and the pathology of oppression it generates, implicitly rejecting the idea that human suffering is a result of some lack of natural bounty. If this point is accepted, it would then be far more realistic to advocate for universal creature contentment given that it's not nature's productivity but human and social distribution that is the issue for Brecht. Because contemporary culture has organized itself in a manner that prevents sufficiency to many while it allows plenty to the few, it should be within human grasp to eliminate such inequities. In turn, current laws of behavior and codes of morality that work to distract the suffering many from their physical torment will change once material conditions improve.

Logical though the above may be, logic alone will not transform the human condition. Recall that, just as occurs in *The Beggar's Opera*, in *The Threepenny Opera* Macheath is saved from the gallows at the last moment. But this time, instead of the "authors" of the play simply rewriting the play's end, Macheath here receives a *royal* reprieve, pension, and title. It seems that we should conclude that Macheath's elevation to the peerage means that the lords of the realm recognize qualities of selfish preservationism in Macheath that they admire, implying, of course, that they are themselves no better than criminals. Given that Macheath displays no other virtue than an overriding, personally rapacious, sexual urge bent primarily on self-indulgence, admitting him into the ranks of human aristocracy reminds us by indirection that the Marxist agenda of equitable redistribution of natural bounty seems far from human reach. Personal advancement will trump a larger communal consideration almost every time.

Brecht can argue for as long as he chooses that humanity must change its ways and embrace a cooperative economics. But such an affirmative, basically legislative agenda will likely not work as long as there remains an embedded, culturally individuating preference to overcome and overpower others in a dash for creature comforts. Serving this individualized end can be quite tantalizing, especially if personally successful. As noted earlier, humanity's survivalist tendencies are determined to preserve the individual and not the species. This, of course, is the point that all rationalist dreamers fail to recognize, believing as they do that to "tell" humanity to do this or that "for the good of the species" will somehow succeed. This is a point implicit in Bakhtin's carnivalesque idealism, where he sees the human species basically as one massive, pulsing body. And it inheres in Brechtian Marxism as well. For the good of the species, we *should* all cooperate. But what is it in this invocation that assures the continued good of "me," especially if I am in a position of privilege, reaping the benefits of oppressions? This matter is what Critchley implies when he warns of over-romanticizing the body. It can be a voracious organism, and its aggressive tendencies need control, which is likely why comic movements often sober up by play's end and why initially unadulterated comic self-indulgence sometimes takes a sort of reactionary, even Puritan, turn. We may in fact be "not good enough just yet," as Brecht would say.

But activating agelastic restraints designed to control overwhelming, indulgent self-interest is not necessary. There is another option that argues against stifling, agelastic restrictions. Buried in Brecht's play—and in Shakespeare's and Jonson's and Rabelais' works—is another alternative proffered quite articulately by contemporary cognitive science that salvages (or reveals) a comic strategy frequently—and regrettably—underutilized and often overlooked.

Coming to Terms with the Gravitational Pull of Mental Materiality

Brecht's famous attempt at *Verfremdungseffekt*, sweeping transformative emotive responses from his theatre in favor of affirming rationalist clarity, has regularly been undermined by what appears to have been the unintended overpowering emotive effect of his plays, *The Threepenny Opera* certainly included. The rich emotive engagement of Brecht's theatre, as well as Shakespeare's, Jonson's, and Gay's, clearly argues that theatre in general and comedy in particular are not places for logical, mental, rationalist activity. In the best of situations, neither can theatre contain itself to the sardonic smile that Meredith champions, overflowing as comedy does into visceral responses beyond the control of conscious awareness.

Of course, coolly looking down from the grandstands, as O'Neill's old foolosopher attempts to do in *The Iceman Cometh*, does involve a tantalizing level of rationalist disentanglement, which would allow us to observe and pass judgment on sloppy, chaotic humanity and determine what exactly it is that we want to call right thought or right action. And diving into the carnival of living does seem to be an indulgence to be avoided because, well, it's just not right. Plus, it clouds our better judgement. Or so goes the agelastic argument of our culture's power brokers. And frequently comic artists allow such somber reflections to infest their works as well.

But we are creatures of flesh and bone, and this is not necessarily a demeaning observation somehow reducing human status to barely above the beasts of the field. This observation begs that we avoid the nearly universal human urge to see our involvement in the physical world as something less than human. As noted earlier, recent breakthroughs in cognitive science suggest that humanity should not only passively tolerate but also actively indulge its carnal inclinations in order to develop a more vital, rounded humanity. But, significantly, the process entails more than merely pursuing and satisfying creature needs, which would indeed be an errant "over-romanticization" of the body.

Antonio Damasio digs back to the seventeenth-century Dutch philosopher Baruch Spinoza (1632–77) to find an explanation of the interactions between mind and body that current neuroscience is today confirming: "Spinoza's solution hinges on the mind's power over the emotional process, which in turn depends on a discovery of negative emotions, and on knowledge of the mechanics of emotion. . . . Freedom is one of the results, not of the kind usually contemplated in discussions of free will, but something far more radical: a reduction of dependencies on the object-emotional needs that enslave us. Another result is that we intuit the essences of the human condition. The intuition is commingled with a serene feeling whose ingredients include pleasure, joy, delight."[4] The crucial idea here is that Spinoza is suggesting a feedback loop wherein the mind reacts to the body and the body reacts to the mind. This effort to know, as Spinoza formulates it, clearly must involve more than isolated and idealized mental activity and returns, post-Descartes, to something more akin to humoral physiology. We must look to understand distinctions between good and bad emotions, positive and negative experiences with the world in which we're embodied. In this regard, Damasio observes the basic understanding put forward by contemporary cognitive sciences in general: "[I]t is apparent that we must rely on highly evolved genetically based biological mechanisms, as well as on suprainstinctual survival strategies that have developed in society, are transmitted by culture, and require, for their application, consciousness, reasoned deliberation, and willpower."[5] This is something that "Western and Eastern thinkers, religious and not, have been aware of . . . for millennia; closer to us, the topic preoccupied both Descartes and Freud, to name but two" (124).

Regarding Descartes' assessment that the "lower" instinctual faculty needs rational control by a human act of will, Damasio agrees, "except that where he specified a control achieved by a nonphysical agent [the mind] I envision a biological operation structured within the human organism and not one bit less complex, admirable, or sublime" (124). And although Freud tried to escape Cartesian dualism by studying the interconnecting superego, he failed to engage specifically biological operations in his study. What needs to be accepted and understood is the full degree to which mind is embodied in the corporeal brain and therefore affected by the biologically physical.

Involvement and engagement of the body with its environment, and in turn of the body with its mind, preclude the objective enterprise of prior philosophical undertakings (Plato's and Descartes' certainly included) and encourage the very undertakings of Damasio and his colleagues. Consciousness theorist Daniel C. Dennett observes, "Today we talk about our conscious decisions and unconscious habits, about conscious experiences we enjoy (in contrast to, say, automatic cash machines, which have no experiences)—but we are no longer quite sure we know what we mean when we say these things."[6] Behavior is generated at so many levels within our neural systems, inter-reacting up and down the scale of neural sophistication, that the idea of objective, disinterested behavior—and even disinterested thought—seems not only an impossible goal but also one not really worth attempting.

The proposition that disinterested thought is not worth attempting is supported by Damasio, who argues that "[e]motions and related phenomena are the foundation for feelings, the mental events that form the bedrock of our minds and whose nature we wish to elucidate."[7] Survival can occur at the instinctual, nonthinking (non-self-aware) level, but the chances of surviving in changing and changeable environments and the possibility of flourishing rather than merely surviving are improved when higher orders of thought creatively interact with the more basic levels of response behavior. Nowhere in this model is there a place for mental processes separated from their bodily connections. In fact, the mind isolated from the body becomes a menace to the now othered body. Clearly Hamlet's mind menaced Hamlet's body.

Hamlet's painful (re)turn to his body is a human reminder that separation threatens homeostasis—the balance of life—and exposes the body to conditions inconducive to survival. Tragedy, by negation, reminds us of homeostatic necessity. Comedy, by celebration, offers the same message.

This all likely sounds quaintly abstract, "philosophical," and irrelevant to our daily lives, perhaps in the same way that comedy does. But Judith Butler summarizes exactly how this all has serious ethical import and, more importantly, exactly how such matters of materialization are of dire survivalist consequence as well. Observing that Plato identifies the male/masculine as the most materially uncompromised possessor of higher reasoning capabilities (the least materially corrupted of all possible candidates) and that "he" is therefore most justifiably master of his material domain, Butler summarizes,

> Plato's scenography of intelligibility depends on the exclusion of women, slaves, children, and animals, where slaves are characterized as those who do not speak his language, and who, in not speaking his language, are considered diminished in their capacity for reason. This xenophobic exclusion operates through the production of racialized Others, and those whose "natures" are considered less rational by virtue of their appointed task in the process of laboring to reproduce the conditions of private life. This domain of the less than rational human bounds the figure of human reason, producing that "man" as one who is without a childhood; is not a primate and so is relieved of the necessity of eating, defecating, living and dying; one who is not a slave, but always a property holder.[8]

If carnival and comedy can bring "man" down from his objective and objectifying high horse, then the beginnings of a very serious transformation can begin, one that could actually transform human interactivity and generate a more completely just, ethical, and equitable existence than has previously existed, though it has often been dreamt of. This, at its heart, is what comedy works to undertake.

So, clearly, comedy needs to be more than a mindlessly carnivalesque celebration of the body or of the material accumulations of our mortal lives. Some comedy remains within those limits, and frequently it is the case that even when comedy does rise above or beyond these levels, what is most remembered is the celebration of carnality, of a Falstaff over a Hal and of a life unfettered rather than an integrating transformation well considered. This is what's "fun" in comedy. But to truly influence existence, comedy must be more than just fun.

Calling an End to Sophistic Idealism: Aristophanes

Erich Segal's discussion of Aristophanes reminds us, too, that fun—or perhaps indignation, depending upon one's inclination—is what we generally recall from comedy: "To many readers, Aristophanes is still synonymous with 'obscenity.' To these over-simplifiers, the epithet 'Aristophanic,' usually interchangeable with 'Rabelaisian,' merely denotes coarse humor, addenda to pudenda, and the comic cacophony of belches and slaps with a stick."[9] Taken alone, "Aristophanic/Rabelaisian" humor, so described, plays an indisputably significant role in the comic agenda, opening up our prudishly puckered bodily and social orifices and calling into question who or what exactly is in charge. The humor provides the opportunity both to remind us of our inescapable attachment to our bodies and to help us discover how these bodies operate as crucial governing agents, the abandonment of which leads invariably to tragic circumstances or at very least aggressive oppression and a master-slave world of sorrow and worry. Segal continues, still speaking of Aristophanes: "True, his characters speak in *le langage vert*, as the French refer to the four-letter words which titillate an audience (colorfully transmuted as 'blue' words in English). But if this were all there was to it, it would be no reason to rank him as a classic" (44). "Classic" or not, however, "blue" words and their attendant actions are nonetheless important human attractors. In Aristophanes' well-known *Lysistrata*, for example, the title character points out exactly how the blue life attracts:

> Announce a debauch in honor of Bacchos,
> a spree for Pan, some footling fertility fieldday,
> and traffic stops—the streets are absolutely clogged
> with frantic females banging on tambourines. No urging
> for an orgy![10]

People are clearly and naturally drawn to the call for "play." And in this particular case, Lysistrata's call for a "serious" meeting leads to an innuendo-ridden exchange, including the hope precisely for the kind of "blue" gathering that would stop traffic:

KLEONIKE: Incidentally, Lysistrata, just why are you calling this meeting. Nothing teeny, I trust?
LYSISTRATA: Immense.
KLEONIKE: Hmmm. And pressing?
LYSISTRATA: Unthinkably tense.
KLEONIKE: Then where IS everybody?
LYSISTRATA: Nothing like that. If it were, we'd already be in session. (351)

The meeting is held, and the "serious" complaint is registered that the women's men are at war and are fulfilling their martial rather than marital obligations. Kleonike agrees, and adds,

> And lovers can't be had for love or money,
> not even synthetics. Why, since those beastly Milesians
> revolted and cut off the leather trade, that handy
> do-it-yourself kit's *vanished* from the open market. (359)

The war of the sexes, innuendo and more, has begun.

We are drawn to the play because of its bawdy humor. But the bawdiness opens both the oppressive masculine on stage and the oppressive masculinized cultural presumptions in the audience (it is hoped), which leads in turn to a prediscursive assessment of the imbecilic consequences of male idealized aggression when compared to the feminized material perspective of coexistent cooperation. That which the Platonic masculine holds in suspicion—everything material and (therefore) feminine—could quite literally save Greece from the devastations that are the Peloponnesian Wars. The idealized masculine and its urge to dominate, of course, historically predominated to the eventual catastrophic detriment of Athens and Sparta alike. If only they had listened to their bodies. If only they had heeded the pleas of the females they had reduced to lesser-reasoning others. The message begins with sexual overflow and scatological indulgence, but it leads to the dire warning that ignoring or avoiding either or both can have (and has had) tragic consequences.

Generally speaking, Aristophanes' plays are full of sexual overabundance. "Rabelaisian" scatology pervades as well: consider the opening scene of *Peace* where two servants are kneading dung cakes. Funny stuff, and memorable, too, but while lesser comic practitioners have made careers on such shenanigans—effectively opening at least one orifice through laughter to receive the odorous material world "as it is"—such an agenda alone is incomplete. There is some value to this agenda, true enough, but there must be more. And while Bakhtin decries the modern inability fully to grasp the importance of Rabelais, it may not be the case that we fail to grasp it as much as it is a case that perhaps there needs to be "more to it" than unstructured random Rabelaisian or Aristophanic liberty.

Though *Lysistrata* is a telling exception, Aristophanes' comedies invariably center around rejuvenating an old man in ways that strike the modern audience as curious, given that comedy typically celebrates the triumph of youth and vitality over age and agelism. This typical youth-defeats-age celebration is the signature of the Bakhtinian and Rabelaisian notion of comic process, but it's not Aristophanes' notion, at least not in and of itself. In *The*

Acharnians, for example, Dicaeopolis (Segal notes it is a charactonym meaning "John Q. Righteous City" [48]) demonstrates widespread disgust with his fellow Athenians (all libertines or cowards) and decides to sidestep the ravages of the ongoing Peloponnesian War by independently brokering a peace with Sparta. Successful, he then goes to a rustic setting (Acharnia) outside of Athens, which is holding a festival, to announce his personal negotiating triumph. Celebrating his accomplishment with full, blatant Dionysian fervor, he becomes the envy of embittered and impotent old men, victims of the ravages of war that have sterilized the world. Eventually, Dicaeopolis succeeds at convincing half the populace of the value of suing for peace. The other half violently opposes Dicaeopolis and his converts. Out of that crowd rises Lamachas, ready to do battle with Dicaeopolis, and the two engage in a duel of phallic slapstick spiced with exchanges of verbal obscenities.

By play's end, Dicaeopolis becomes a victor on all fields, martial and amorous, the focal point of general and universal reinvigoration wherein the ravages of time have been undone. As Segal notes, although "in all later comic genres the vigorous phallus at the end belongs to the younger generation . . . here in Aristophanes it belongs to the newly young *re*-generation" (57). Instead of handing off to another generation the reins of progress, enlightenment, and hope, here the same biological entity *re*generates into an organism exhibiting personal vitality in a self-contained cycle of rebirth. Unlike the traditional pattern, wherein death and decay claim one generation and prepare the way for the next generation, in Aristophanes we have the hint that it is not necessary for life actually to end in order for renewal to edge its way into existence. Rather, we have the suggestion that biology can actually self-regenerate, curiously without the instrument of painful death, and that time in some seemingly impossible way can be reversed or, more possibly, can be imbued with a value-added quantum leap of progress over or beyond the limitations imposed by a linearity whose end product is death. Unlike the traditional reading of nature that suggests winter is death and spring rebirth, Aristophanes suggests that the single organism—nature/earth/human—reenergizes after mere dormancy and reawakens to a renewed youthfulness.

This is Dicaeopolis, superannuation rejuvenated. So while it is generally the case that death and rebirth are how winter to spring is envisioned, Aristophanes presents us with an even more positive viewing that erases death from the picture. Indeed, in comedy, the "death" phase is invariably metaphorical, more directly a case of dormancy as Aristophanes proposes than actual death. It's an ethos best located in the country logic of agrarian economics, in that world whose vital activities precede and then lead up to that other carnivalesque opportunity, the late-summer marketplace of harvest. It is the autumnal, agrarian conclusion to that initial springtime awakening where the ethic of perpetuating, bountiful life strives to infiltrate orthodox cultural belief systems.

This brings us to a fundamental distinction between tragedy and comedy. Tragedy requires death for regeneration (with very few exceptions), while the path of comedy need not pass through that unknown country in order to arrive at virtually the same regenerated destination. From an orthodox, masculinized perspective, life can only be affirmed through the "gift of death" since the masculine cannot generate life in any literal sense and in fact must avoid affiliations with such natural conceptions of material (re)generation if *he* is to remain an idealized objectivity. So life can only be masculinely confirmed and controlled by appealing to death, sometimes to the point that it calls for generating a sacrifice even against orthodox ethical standards. Derrida, among others, looks especially at the story of Abraham and Isaac (following Kierkegaard), wherein Abraham must sacrifice the son he most prizes, against all logic, in order to prove his love of God.[11] However, God converts this gift of death, heroically/tragically proffered, to a gift of life by staying Abraham's deadly masculine/fatherly hand. The result, if anything, is that the near-tragic act of sacrifice becomes a comedy of reversal and a gift of good fortune, a celebration of life that confirms faith (the new order) over reason (the old order) and creates a promise of a new world governed by grace. The whole story of Abraham and Isaac is fundamentally comic.

What we see here (though neither Derrida nor Kierkegaard touches on the point) is what comedy in general does: it confronts masculine aggression and the ethos of sacrifice and death, replacing them with the gift of grace derived from "feminized" nurture, bounty, and life, precisely what orthodox culture cannot permit without abandoning its idealized cornerstones of civilization founded on control of materiality. This is a crucial point to keep in mind: comedy sees life to be not so much dependent on coping with the solipsizing inevitability of death, as it more generally welcomes a nurturing focus on life *before* death or sometimes even *without* death, seeing existence as a continuing procedure akin to "molting," leaving behind an old pelt and donning a vitally fresh, new skin. This perspective bypasses the obsessive recognition of individuating death or mortality that drives the former worldview. It's a perspective that generally moves up one level to the life-continuing flow of individual genetic *transfer*, best articulated by Benedick's famous expostulation, "The world must be peopled" (*Much Ado About Nothing*, II.iii.237).

Even in the early life of comedy, these points of material, agrarian truth were understood and juxtaposed against an already orthodox opposition. Take Aristophanes' masterpiece *The Clouds*. It introduces a commonsensical country hero much like Dicaeopolis but one who unfortunately succumbs to the city's urbane, sophistic urges to escape the country wisdom upon which the Greeks had once so successfully thrived. The Athenian decision during the Peloponnesian Wars with Sparta to sacrifice the countryside in order to preserve the city is Aristophanes' point of criticism, a reasonable one for *any* comic artist given that Athens gave up its agrarian roots to preserve an increasingly unsavory, overcrowded metropolitan alternative. The loss of contact with the living matter of the fields swings us into the project of decrying the loss of our materiality. So Aristophanes' literal, limited satiric assault on his contemporary situation assumes a larger comic value in the more general sense (as does most of Aristophanes' work, *Lysistrata* included), becoming lastingly comic even if it originated as a timely satire. The very idea that urban "sophistication" is privileged over grounded, rustic wisdom remains the central theme of the strongest of comic works. Here Aristophanes has located clear, particularized, historical groundings to actualize the comedy agenda. And while farmers are generally considered to be an entrenched conservative breed, perhaps their wisdom actually is, ironically, the wisdom of revolution, comic or otherwise.

The hero of *The Clouds*, Strepsiades is a rustic hero drawn into the temptations of the city, having married a demanding city wife and raised a playboy son who are draining the resources of his country estate. His situation echoes the situation of Athens proper: the city has pursued an imperialist agenda leading to a war that has resulted in sacrificing the countryside in order to preserve the city. From this point, as William Arrowsmith observes, "Cut off from the earth and uprooted from the context that gave it life and value, the Old Order had decayed, and with it were being destroyed all those traditions and virtues and decencies, which, for a conservative countryman like Aristophanes, were synonymous with Athenian civilization itself."[12] Strepsiades is the long-suffering rustic longing for the good old days. Musing beside his sleeping son, he reminisces,

> I used to be a farmer—the sweetest life on earth,
> a lovely, moldy, unspruce, litter-jumble life,
> bursting with honeybees, bloated with sheep and olives.
> And then, poor hick, what did I do but marry
> your mother, a city girl, and niece of that Megakles
>
> She was a pretty niece: Miss Megakles-de-luxe
> Well, so we got married and we clambered into bed—
> me, a stink of wine-lees, fig-boxes, and wool-fat;
> she, the whiff of spices, pure saffron, tonguekisses,
> Luxury, High Prices, gourmandizing, goddess Lechery,
> and every little elf, imp, and sprite of Intercourse.[13]

Suffering an insatiable wife and arrogant spendthrift of a son, Strepsiades plans to allay his impending bankruptcy by enrolling his son in "Sokrates's [sic] Thinkery," which has become all the fashion and has led to household gods being replaced by a potbellied stove representing the "MODEL OF THE UNIVERSE ACCORDING TO THE CONVECTION PRINCIPLE" (22). This newly abstracted universal principle, derived through rationalist abstraction, will teach his son "that the whole atmosphere is actually a Cosmical Oven and we're not really people but little pieces of charcoal blazing away" (29). Additionally, and more importantly, the Thinkery offers a course utilizing its abstracted rationalist processes, entitled "The Technique of Winning Lawsuits" (29). That course, Strepsiades hopes, will save him from financial ruin.

But his son refuses to sign up, so as a last resort Strepsiades signs up himself, learning instantly the genius of the Thinkery when he discovers that his knocking at the door miscarries an experiment designed to measure the distance a flea can jump. He also learns that the Thinkery is hard at work determining the alimentary processes of a gnat, which inspires Strepsiades to proclaim, "Why, the man who has mastered the ass of the gnat could win an acquittal from any court" (33). Sokrates is revealed suspended in a basket staring at the heavens while emaciated students are "*deeply engaged in a rapt contemplation of the ground*" (35). Other students are bent over studying Hades, "their asses scanning the sky" because they are "[t]aking a minor in astronomy" (36). If the absurdity isn't evident by this point, additional asininities are revealed, but most significantly, Strepsiades learns that in this Thinkery, "[t]he statutes clearly forbid overexposure to fresh air" (36). Hovering above nature itself, Sokrates proclaims,

> You see,
> only by being suspended aloft, by dangling
> my mind in the heavens and mingling my rare thought
> with the ethereal air, could I ever achieve strict
> scientific accuracy in my survey of the vast empyrean.
> Had I pursued my enquiries from down there on the ground,
> my data would be worthless. The earth, you see, pulls down
> the delicate essence of thought to its own gross level. (40)

Believing thought to be of an origin different than physical reality, Sokrates sees the mind as an essence separate from the brain or any other corrupted corporeality. Furthermore, according to Sokrates, the mind itself has replaced the gods: "The gods, my dear simple fellow, are a mere expression coined by vulgar superstition. We frown upon such coinage here" (41). Sokrates' invocation reveals his new god:

> O Lord God Immeasurable Ether, You who envelop the
> world! O translucent Ozone!
> And you, O lightningthundered
> holy clouds
> Great Majesties, arise!
> Reveal yourselves to your Sophist's eyes. (43)

On the arrival of the "lightningthundered" holy clouds, Strepsiades, in typical Aristophanic manner, announces, "Sacrilege or not, I'VE GOT TO CRAP" (45), to which Sokrates retorts, "No more of your smut. Leave filth like that to the comic stage" (45).

Clearly Aristophanes' Sokrates is an advocate of a separation of mind and body wherein abstracting mind exudes purified truth and grounded body bleeds putrefied waste to be avoided at all cost as the human enterprise pursues an understanding of the cosmos. Sokrates observes, "Clouds are . . . patrons of a varied group of gentlemen: chiropractors, prophets, longhairs, quacks, fops, charlatans, fairies, dithyrambic poets, scientists, dandies, astrologers, and other men of leisure. And because all alike, without exception, walk with their heads among the clouds and base their inspiration on the murky Muse, the Clouds support them and feed them" (47). Impressed, Strepsiades pledges allegiance to Sokrates' trinity, "GREAT CHAOS, THE CLOUDS, and BAMBOOZLE" (55), and he is brought into the mystery of the clouds by entering Sokrates' cave. Following a brief interlude, however, Sokrates enters and announces Strapsiades' utter stupidity. Hopelessly literal and endlessly practical ("Will rhythm buy the groceries?" [71]), Strapsiades suffers the indignities of bug bites while he is supposed to be reflecting on his master's lessons. Parodying the Socratic method, Strapsiades asks what he should be contemplating, only to be told by Sokrates, "A moot question, friend, whose answer lies with you. When *you* know what *you* want, kindly illuminate *me*" (79).

Failing in his education, Strapsiades is ordered to bring his son to the Thinkery for training. The reluctant son, Pheidippides, is introduced to Philosophy and Sophistry (in some translations labeled Right Logic and Wrong Logic, respectively),[14] literally caged creatures with bodies of humans and heads of fighting roosters. A verbal battle ensues, Philosophy defending the Old Education of "Homespun Honesty, Plainspeaking, and Truth" and "the regime of the three D's—DISCIPLINE, DECORUM, and DUTY" (97). Sophistry retaliates by acknowledging, "I freely admit that among men of learning I am—somewhat pejoratively—dubbed the Sophistic, or Immoral, Logic." However,

this little invention of mine,
this knack of taking what might appear to be the worse argument
and nonetheless winning my case, has, I might add, proved to be
an extremely lucrative source of income. (103)

Sophistry proceeds to defeat Philosophy on all fronts. He even successfully argues that being reamed up the rectum with a radish would *not* be a disgraceful thing. Reluctantly conceding that the world is populated by "Buggers," Philosophy tosses his cloak and asks to be welcomed as a Bugger himself. Sophistry wins. And Strepsiades wants his son to become a Sophist in the bargain, in order, discreditably, to defend his estate against lawful creditors. All seems fair when one dispenses with using a well-calibrated moral compass and turns to greed as a guide.

Emerging as a trained Sophist, Pheidippides brilliantly confuses the issue of payment "Dueday" and impresses his father into believing he's now invincible to all lawsuits. After a brief spell of sensed triumph, however, Strepsiades comes on stage having been beaten by his son, who proceeds through Sophistry to prove it was a justified action. Corporally punishing children is a parent's proof of love for that child; therefore, "lickings" and love being synonymous. Sophistically speaking, then, it is actually a son's duty to beat his father. Having been beaten by his son, literally and figuratively, Strepsiades finally sees Sophistry for what it is, pleads for forgiveness and direction from his household god Hermes, is commanded to "BURN DOWN THE THINKERY" (143), and in a parody of the pot-bellied stove theory of universal convection, he and his servants burn down the Thinkery, leaving the Chorus to announce,

Now ladies, let us leave
and go our way.
Our dances here are done,
and so's our play. (147)

Strepsiades remains in debt, and his household is still in encumbered disarray, but he is nonetheless rejuvenated into action based on a rustic faith in the bounty of a well-grounded life that destroys his cloud-bound fantasy of painless escape from personal liability. Presumably reassuming his rustic reliance on noncosmopolitan, antisophistic common sense, he will pursue a path to economic solvency grounded on his skills as a guardian of and adherent to country values founded upon an economy of earth and body.

Why Rusticity?

Strepsiades' rusticity is compromised by illusions of a sophistical quick fix. Once the delusion becomes apparent, the Right Logic with which he has created a good life reasserts itself. In contrast to this simple man living a life of more fully embodied, "natural" order stands Sokrates/Socrates (whom Aristophanes may or may not have been directly attacking), a man determined to ignore the ways of the world in favor of ideals of the mind. "Right logic" is naturally founded good sense. What we have is the common sense of homegrown, land-based consciousness falling under the spell of "objective," abstracting rationalism capable of drawing any number of conclusions and unencumbered by material, physical evidence except by way of a certain selectivity designed to support a desired "idealized" conclusion. The mind works in isolation and stands superior to the body.

Significantly in *The Clouds*, as Sokrates the idealist sophist is satirized, so is Euripides the idealizing tragedian. (He's an object of derision in *The Acharnians* as well.) Euripides is portrayed as a man dreamily and airily working without connection to the real world while writing his plays.[15] The connection between idealism and tragedy is, of course, more than mere coincidence given that, according to Aristophanes at least, these "thinkers" lose touch with the life force that propels the lesser figures Strepsiades and Diceaopolis to sounder, more solid conclusions. Comic rusticity originates in the embodied, enrooted, and even scatologically endunged soil and develops a sense of interdependency espoused by advocates of liberation, equity, and justice. On the other hand, any masculinized idealist of autonomous self-control will reject the kind of interdependence on materiality that is the cornerstone of the agrarian way of life since masculinity and orthodoxy sees (inter)dependence as loss, subjection, and enslavement. But as Kelly Oliver observes, "Why does dependency have to be figured as violent, alienating, subjugating, and dominating? Only if we start with the ideal of the self-possessed autonomous subject is dependence threatening. If, however, we give up that ideal and operate in the world with a truly interrelational conception of subjectivity, a subjectivity without subjects, then dependence is seen as the force of life, as the very possibility of change, rather than as the paradoxical life bought at the expense of violence and death."[16] Interrelationality is what comedy's ally, agrarian rusticity, strives to reveal. Perhaps, contrary to popular opinion, the most radical notions actually derive from the land, the material, and the comic rustic closest to the source.

Aristophanes' rustics are the beginning of a long tradition in comedy, found, for example, in the creation of Cervantes' Sancho Panza, Fielding's Tom Jones, and Twain's Huck Finn. Like Strepsiades, these "naturals" can fall under the spell and even look ridiculous when exposed to civilization's cosmopolitan, urbane trappings. Sometimes they fall to glittering culture's tantalizing offerings, as in the case of Falstaff, who works so hard to beat a system determined to beat him. Other times they simply decide to "go to hell" and then "light out" altogether, as is the case of Huck Finn, who decides not to accept the "civilized" sophistry that endorses aggressive dominion and slavery. Regardless of ultimate outcome—though it's always preferable that they prevail—their lust for life holds sway over the stage.

It must be emphasized in the final analysis that for comedy "seriously" to complete its task, it must give us more than a band of memorable roustabouts pursuing egocentric paths of self-gratification, no matter the amount of pleasure derived from witnessing and even dreaming about such indulgences. At this level, we return to Damasio's acknowledgement of Spinoza's key addition to the value of understanding existence by understanding our physical conditions. What is necessary is "a virtuous life assisted by a socio-political system whose laws help the individual with the task of being fair and charitable to others—but then it goes further. Spinoza asks for an acceptance of natural events as necessary, in keeping with scientific understanding."[17] Descartes' struggle was a continuation of Socrates' agenda, namely to escape the earthbound realm of sloppy existence in the pursuit of objective, rationalist truth. Spinoza saw the need to ground that thinking into the earth itself even while accepting a need for consciousness to feed back into that reservoir of truth and reality. Unconsidered, untempered, and ungrounded behavior would be intolerable in the final analysis. Falstaff is an excellent case study of exuberantly enthusiastic irresponsibility that draws its audience in, but that ultimately isn't enough. We love to see it, wish we were as unfettered, and wish there were a world where such indulgent exuberance is the *norm*. It seems so *natural*, after all. But then we must recall that the exuberance we hold up as so *natural* and that is so constrained by social mandates may not be so natural after all. After all, unrestrained exuberance leads to self-interested aggression, which even in a state of nature must have its limits. At best, such exuberance exists under certain conditions only temporarily, in places and at times where its cataclysmic potential finds release without indiscriminately destructive results. It can occur in Eastcheap, but only briefly and presumably not in Windsor. However, Eastcheap can influence Windsor as Falstaff influenced Hal. And that's important. But also important is the need to recognize that unchecked exuberance is not *really* a state of nature any more than its mortal foe agelasm is. Having developed our own tools of mortal aggression through our long course of ingenious mental development, and not having been naturally programmed to redirect that mortal potential as other aggressive animals have been programmed (in conjunction with their evolved tools of death, like teeth and claws), our natural aggression has outrun our native instincts to control its mortal potential. Exuberance and its more threatening relative, aggression, must be controlled naturally by animal instinct or by more self-conscious means. Often lacking the more instinctive controls, the human animal must utilize those more self-conscious options.

This gets us back to *The Threepenny Opera*. Any number of philosophical and political treatises have been published utilizing rational discourse to argue the Marxist agenda of equitable distribution of capital, some rather brilliantly worthy of Sokrates' Thinkery. What Brecht's Gay adaptation offers, however, scrapes at our consciousnesses from beneath and around our rational, abstracting sensibilities. The agenda is the same, but the means of delivery profoundly different. Through embodied comic strategy, Brecht's *The Threepenny Opera* effectively conveys the transformative need for capital redistribution by appealing directly not to the disembodied mind through Marxist logic, but to the mind through the appetitive urges of the comic, carnivalized body. What the process suggests is that our urges may be "good" but that controls are necessary in a human race that has outrun its genes. And dreams of unchecked exuberance as somehow natural have allowed for aggression that has tilted our world in favor of the fortunate few.

Turning our bodies into better-tuned receptors may result in actually hearing new ways of being by circling around or burrowing beneath the currently over-present resistances our agelastic impositions place before them. Returning to some romanticized state of nature will not work. Learning to tune in to our bodies, however, *is* a crucial first step to rebalancing our world of freedom and control.

CHAPTER 5

The Orderly Disorder of Comic Vitality and Its Liberating Potential

There is something ringingly true about Nietzsche's opening words to *The Birth of Tragedy*: "Much will have been gained for esthetics once we have succeeded in apprehending directly—rather than merely *ascertaining*—that art owes its continuous evolution to the Apollonian-Dionysiac duality, even as the propagation of the species depends on the duality of the sexes, their constant conflicts and periodic acts of reconciliation."[1] Seeing a biological parallel to the art he describes, Nietzsche goes a step further: "I have borrowed my adjectives from the Greeks, who developed their mystical doctrines of art through plausible *embodiments*, not through purely conceptual means" (19). What Nietzsche identifies and what the Greeks developed (if Nietzsche is correct) derived from direct experience, and, furthermore, what the Greeks presented in their art required direct apprehension rather than abstract ascertainment.

Put another way, Aristophanes' Sokrates and Thinkery are the sterile off-spring of an epistemological enterprise gone off track. Nietzsche's agenda of focusing on the "other" nonidealist/non-Sophistic Greeks rather than the conceptual idealists is an agenda paralleling Damasio's materialist cognitive science, anticipating it even as Rabelais' art and humoral physiology anticipate it. As such, comedy rather than tragedy may have been where Nietzsche should have mined for material as he pursued his "gay science" that celebrated present life rather than timeless abstractions. As Sypher points out, unlike the tragedian, "the comic artist begins by accepting the absurd, the 'improbable,' in human experience."[2] Closure, stasis, and even causal likelihood are not the norm in the comic vision because they are frequently not the norm in nature, and so the nearly unlimited horizon of possibility is the comic domain, the very domain Nietzsche seemed most interested in championing.

That said, it should nonetheless be noted that Nietzsche's fusion of disorderly Dionysian and orderly Apollonian urges helps also to clarify a suspicion that Aristophanic "Right Logic" rings a bit too much like "proper" thinking or "correct" behavior, following conservative, agelastic satirism rather than a comically exuberant line of thinking. But in a twist that reverses what is generally thought to be Right Logic, Aristophanes sees it as a "logic" grounded in natural paradigms rather than rigorously rationalist formulations of propriety. Sypher reminds us that the comic artist "has less resistance than the tragic artist to representing what seems incoherent and inexplicable, and thus lowers the threshold of artistic perception. After all, comedy, not tragedy, admits the disorderly into the realm of art; the grotesque depends upon an irrational focus" (23). Right Logic from this perspective embraces qualities of thought that are frequently dismissed as patently *un*logical. As Sypher describes the comic through allusion to the Rabelaisian grotesque, so is Nietzsche's assessment of the Dionysiac, implicitly Rabelaisian and carnivalesque, aroused as Nietzsche claims it to be by narcotic potions "or through the powerful approach of spring, which penetrates with joy the whole frame of nature" (22). Nietzsche continues, "So stirred the individual forgets himself completely. It is the same Dionysiac power which in medieval Germany drove ever increasing crowds of people singing and dancing from place to place. . . . There are people who, either from lack of experience or out of sheer stupidity, turn away from such phenomena, and strong in the sense of their own sanity, label them either mockingly or pityingly 'endemic diseases.' These benighted souls have no idea how cadaverous and ghostly their "sanity" appears as the intense throng of Dionysiac revelers sweeps past them" (23). The reveling unites individuals into an undifferentiated mass, breaking down individualized fear of death in celebration of the immortal "life" of the social unit. But also, "nature itself, long alienated or subjugated, rises again to celebrate the reconciliation with her prodigal son, man" (23). If Strepsiades is hoping for a "miracle," he is more likely to find it in his agrarian setting—working for a bumper crop of bounty—rather than in the sophistically duplicitous Sokratic Thinkery.

To bottle this hope-springs-eternal spirit of vitality may at first glance appear a fool's venture, but Aristophanic Right Logic, full of an awareness of nature's ripe vitality and potential for overabundance, does what it can to expect the unexpected, to *not* be surprised by what may arise from nature in the world. Sometimes, against all logic and reason, the cavalry does come to the rescue, and solutions do appear out of thin air. Is pointing out such possibilities merely foolish dreaming, what reason will surely tell us is banking against the odds? Or, perhaps, are nature's patterns themselves odds-busters? Perhaps the logic of cynicism is the first thing humanity will be able to minimize if it returns to something more like its natural fold.

The problem here seems to be that we've misinterpreted the apparent fact that nature is "red in tooth and claw" (to quote Tennyson's "In Memoriam"), conceding that so too must be *our* nature. But this dog-eat-dog vision of life has that other side to it, that instinctive quality of redirected aggression that can be located in much of the animal kingdom. This biological phenomenon, in turn, can be seen as an integrative part of a systemic feature in nature that humanity has all too frequently ignored. This more cooperative perspective has been sighted more and more frequently in the contemporary sciences, which are beginning to realize more and more that the apparent randomness against which we constantly struggle is actually beneficial and necessary to existence. Disorder is not something necessarily to be overcome, hemmed in, or eradicated. It is not some natural antagonist hovering menacingly above humanity. Rather, in the best of worlds—literally—order and disorder (like Apollonian and Dionysiac urges) must interact. By extension, carnival and disorder are not "mere" diversions to be cleaned up as early as possible so that we can get on with the real business of orderly existence. Rather, out of disorder (and carnival) comes an "order" different than the agelast's order, called living. It is something that only *appears* to be disordered and only *appears* to require orderly constrains.

In *Science, Order, and Creativity* (1987), David Bohm and F. David Peat make the following observations about these rather recent scientific advances, which generally fall under the popular heading of chaos theory: "Randomness is being treated not as something incommensurate with order but as a special case of a more general notion of order, in this case of orders of infinite degree. This may appear as a curious step to take, since chance and randomness are generally thought of as being equal to total *dis*order (the absence of any order at all). . . . But here it is proposed that whatever happens must take place in *some* order so that the notion of a 'total lack of order' has no real meaning."[3] Order and disorder do not operate in dialectic opposition to each other, nor is disorder the "total lack of order" previously suggested by agelastic apologists of static, hierarchic order. Rather, order-disorder is a continuum found in nature wherein one should consider randomness/disorder to be, as Bohm and Peat suggest, "a limiting case of order," a concept that makes it "possible to bring together the notions of strict determinism and chance (i.e., randomness) as processes that are opposite ends of the general spectrum of order" (132). The interaction between these "ends" of the spectrum—and everything in between—is precisely what chaos theory argues occurs in nature: order rises from disorder, and disorder originates in order. This is chaotics at work

It should be apparent at this point that one could replace "chaotics" and "chaos theory" with "comedy." Comedy has for ages advocated nature's integrated role in human existence, demonstrating the folly of disregarding nature's rhythms and patterns. It has also attacked the arrogance of abstract reason when reason somehow actually conceives nature to be orderly, hierarchic, linear through some mistaken, desire-ridden force of will. In other words, whenever disorder/chaos was cast from our ontological consciousness, comedy was there to remind us of the vital necessity of this essential force of creation, creativity, change. But even when chaos/disorder was included in the social fabric, all too often it was admitted as a sort of harmless release valve, permitted in regulated doses so that we may return refreshed and therefore better able to endure the tedium of *real* life. What chaotics confirm, however, is what at least the more complete visions of comedy have argued all along: namely that order and disorder comprise a single unit joined in a dance that generates life itself. Without one or the other, life simply cannot be. It's a message intuitively gathered in comic form for millennia. But it took Tom Stoppard to merge the poet's vision with a "scientific" rigor that until the late twentieth century simply did not exist. Through Stoppard, we see poets and madmen join scientists and technicians in their own curious dance of disorder and order, in turn shedding light on Stoppard's countless intuitively orderly-disorderly comic predecessors.

Stoppard's *Arcadia*: Studying the Humanity of Chaos

Early in his career, Tom Stoppard created but then withdrew a little-known (and still unpublished) work, *Galileo* (1973). A reaction to Brecht's *Life of Galileo*, Stoppard created a shrewd though ultimately miscalculating Galileo who, before his infamous (and Brecht says cowardly) retraction, presents a vision that embraces disorder and dynamics and that very much aligns with a physically open, dynamically *comic* vision in general. First, we see in standard comic form a Galileo with a nearly Falstaffian longing for life, choosing survival over martyrdom. As John Fleming observes, "Whereas Brecht positioned Galileo's eventual recantation as a betrayal of truth, science, and humankind, Stoppard's Galileo views events through a more pragmatic lens, arguing there would be nothing gained through martyrdom. His death would not change anything, and he can better serve humanity by engaging in what scientific research he can."[4] Clearly a man decidedly opposed to idealistic heroism, Galileo's "cowardice" is, as Auden explained regarding Falstaff's cowardice, an unveiling of idealism as empty and ultimately fruitless. Giordano Bruno may have died for a belief and an ideal, but Galileo chooses to live because life is more important than idealism. This, of course, is pretty standard comic-vision fare.

What are significant of Stoppard's "lost" play are Galileo's conclusions about the universe. Speaking against the static and unchanging perfection of Aristotle's hand-me-down vision, Galileo comments,

> I do not understand why perfection should be a state of rest rather than a state of change. I am very fond of this earth. It is not of course perfect, but that which I find noble and admirable in it is all to do with change. The change of a bud to a flower, of a deer feeding to a deer running, the change of a grape to wine, child to man, wood to flame; and the ash is thrown on the soil to help the buds change to flowers again. Alteration, novelty, decay, regeneration—these are not *blemishes* of the world. Who would want a crystal globe? What use is *that* to man as created by god?[5]

At this stage in his career, Stoppard clearly intuits the chaos paradigm as virtually all great comic artists have intuited it. Change is a vital virtue to be embraced rather than stifled. It is precisely what we've seen in Aristophanes, Jonson, Shakespeare, and others.

However, in 1993 Stoppard presented *Arcadia* and stepped beyond comic "intuition" by directly utilizing the chaotics paradigm to articulate the comic vision with a precision only dreamt of in the philosophies of his comic predecessors. *Arcadia* is a play that uses chaos patterning actually to structure the work. As such, the play is the first mainstream theatre product consciously designed to be formally a "chaos" play. Chaos theory, as Stoppard reports through his character Valentine in *Arcadia,* is a theory that revolutionizes our understanding of "[t]he ordinary-sized stuff which is our lives."[6] Unlike some of the science that has evolved in the last hundred years—think quantum physics and theories of relativity—life as we "know" it is the subject of chaos theory. And through life as we know it through the lens of chaos theory is the subject of the play.

Chaotics directly challenges the idealized static vision of the material world as imposed on nature by classic calculations set up by a rigid, agelast concept of reality. *Arcadia* uses chaotics to challenge this still pervasive classical, overdeterministic miscalculation of nature and its effect on our existence. The play is set in a single room but in two centuries, 1809 and the present. The latter, contemporary time period capitalizes on introducing a character, the graduate-student chaotician Valentine who is able to articulate chaos theory in ways that no one in 1809 would actually have been able to do. But with the 1809 world he creates, Stoppard suggests a vital connection— even a kindred spirit—between twentieth-century chaos thought and the eighteenth/nineteenth-century romantic frame of mind, making special note that both fundamentally challenge the rationalist and neoclassically Newto-nianist frame of mind that dominates both worlds.

The central dynamic spokesperson in the 1809 part of the play is Thoma-sina, a young, intellectually curious girl whose youth and gender minimize her exposure to the culturally encrusted "way things are," allowing her free reign to think culturally unthinkable thoughts. The result is that Thomasina studies nature rather than scholars and arrives at conclusions that anticipate what will eventually become known as the second law of thermodynamics, typically regarded as the first full challenge to the triumphant Newtonian-ism of the age because it works against the notion of Newtonian perpetual order by confirming that loss and eventual heat death are cosmic inevitability. Change occurs naturally, though at this stage in scientific thinking, change is seen only to tend toward dissipation and death. The process of heat dis-sipation is irreversible, and there are no known natural laws that allow for regrouping of lost heat to an original, reconcentrated condition. As Thoma-sina rightly concludes, in rather simple yet profound terms, if you stir a bowl of rice pudding and jam, "the spoonful of jam spreads itself round making red trails like the picture of a meteor in my astronomical atlas. But if you stir backward, the jam will not come together again. Indeed, the pudding does not notice and continues to turn pink" (4–5). Thomasina's demonstra-tion is Stoppard's apt illustration of irreversible entropic dissipation. Hence, Thomasina proves Newtonian law to be incomplete at best.

In *Arcadia*, Stoppard suggests that the British romantic movement can be seen as a period that reveals important critical contours in cultural thought and attitudes that are anti-agelastic and proto-chaotically comic in their embrace of natural behavior. But that movement turns to endorsing an incomplete vision, rather gloomily embracing decaying disorder and even the perceived randomness of nature that so terrifies official culture. However, nature is not merely a randomly disordered system requiring either gloomy, Byronic embrace of the natural or our human counter-urge to control/manipulate it back to an acceptable order. This is a point Stoppard makes as he sets up a contrastive paradigm in *Arcadia* between classical, eighteenth-century sculpted gardening styles and the Gothic, picturesque, romantic wildernesses that would overtake the earlier style. The proposed conversion of the grounds at the country estate of Sidley Park is described by Lady Croom, the resistant lady of the manor:

> Here [in a sketch] is the Park as it appears to us now, and here [in another sketch] as it might be when Mr. Noakes has done with it. Where there is the familiar pastoral refinement of an Englishman's garden, here is an eruption of gloomy forest and towering crag, of ruins where there was never a house, of

water dashing against rocks where there was neither spring nor a stone I could not throw the length of a cricket pitch. My hyacinth dell is become a haunt for hobgoblins, my Chinese bridge, which I am assured is superior to the one at Kew, and for all I know at Peking, is usurped by a fallen obelisk overgrown with briars. (12)

Lady Croom longs for the controlling order of nature found in eighteenth-century neoclassical thought, while the new landscaper Mr. Noakes argues, "Irregularity is one of the chiefest principles of the picturesque style" (12) and by extension the chiefest principle of a new way of viewing nature. An embrace of natural irregularity prevails in the picturesque style over the former mode of endorsing control of naturally seeming unregulated disorder and disarray.

Pursuing her own iconoclastic studies, Thomasina glowingly calls Noakes "The Emperor of Irregularity" (85) and is inspired by the picturesque style to develop a "new" geometry designed to capture her iconoclastic musings, which she calls a "New Geometry of Irregular Forms" (43). However, Noakes' purely impressionistic landscaping romanticism misses a key point that Thomasina hits on: nature has a rising order even amid the apparent disorder. Ultimately, there's no need to impose order as Lady Croom desires, nor to embrace disorder as gloomy Byronism does. Rather, there's the hybridized ground that Thomasina is pursuing, uncovering what is today known as fractal functions, a brand of chaotics that reveals rising order implicit in the seeming irregularity or disorder of nature. She proudly announces that, with her new geometry, she "will plot this leaf and deduce its equation" (37). She is announcing what amounts to a fractal-based procedure that would only be formally developed almost two centuries later. What she is about to hit on is an implicit order in nature that actually grows out of disorder. Nature "illogically" reveals an unforeseen order within apparent natural disorder. Stephen Kellert explains fractals as follows: "A jagged coastline provides a useful example of a fractal-like object. Observed from afar, the coastline reveals some peninsulas and bays; on closer examination, smaller juts and coves are seen, and these again reveal jagged borders when surveyed more closely. If we can imagine the coastline so jagged that with each new level of magnification new details of the terrain appear . . . this is a fractal.[7] What is important is that nature throughout these scaled levels reveals an order that only appears to be disorderly because at each scaled level, similar patterns arise. Furthermore, since such structures defy exact duplication, they result in the morphological diversity that ensures rich natural variety, change and, at least for humans, surprise. Add to this the fact that such diversity does not reduce to mere randomness, for even as these natural processes produce diversity, they remain within certain "ordered" bounds of self-similarity. No snowflake is the same, though all have convergent similarities. So it is with most of nature's products from, to take one example, river and tributary systems to the veins in an oak leaf, to the circulatory system in animals. Up and down the scales of size and even apparent levels of complexity, self-similar structures and patterns recur throughout nature, belying the idea that nature's rich diversity is randomly generated. Orderly disorder rules the day.

From Thomasina's perspective, traditional nineteenth-century thinking in general and mathematics in particular reduces existence to formulae conformable to human/cultural desire, forcing nature to be seen as imperfect shadows of an "ideal" model. Mountains are like giant triangles, trees are like ornamental bulbs attached to vertical posts. In landscape architecture, orderly eighteenth-century practice entails reducing nature so that it conforms to such limited/limiting exactitude of existing formulae. Long sweeping curves and controlled planting are the rule. The picturesque/romantic style, however, encourages breaching the boundaries of those prescriptions and therefore encourages nonconformist minds like Thomasina's to uncover the "logic" that organically describes the naturally occurring irregularities that have thus far been prescriptively ignored as so much unsavory sloppiness.

That even twentieth- and twenty-first-century minds still abide by culturally endorsed Newtonian configurations is verified by the separate endeavors of Hannah and Bernard, inhabitants of the play's contemporary set. Hannah uses her mind to try to prove that the redesigned, picturesque Sidley Park is paradigmatic of the nineteenth-century "decline from thinking to feeling" (27), intimating a desire to use orderly rationalist thought to prove itself superior to lowly picturesque/romantic "feeling," which she describes as the irregularity of sentiment. And Bernard's enterprise is to use his rational, investigative skills to piece together evidence that will prove Lord Byron was involved in a deadly 1809 duel at Sidley Park, solving the mystery of Byron's self-imposed exile thus far unexplained by historians. The process they both utilize reveals a tendency to *manipulate* evidence to prove a point, symptom of urges to control rather than to embrace the irregular world around them. In the process of personal awakening—by way of comic "humiliation"—Hannah learns the value of "feeling," and Bernard learns that a "logical" reconstruction of the past reveals unforeseen trackings on his graph of reconstruction. What their minds have been configured by their dominant-culture educations to expect is fractally undermined by nature and reality itself. At different levels, they both learn something of the value of chaos, of the truth that unexpected order rises out of moments of unpredictability. In the end, it's comic revelation at work. Stoppard's play focuses here on the confused humanity of his characters, the fact that they are *not* rationally in control of their lives as they publicly try to verify.

But Stoppard goes a notable step further. What he's shown thus far is that order reveals itself in apparent disorder. But to this point, disorder still appears to be the eventual result of nature, given its dissipative inevitability. Stoppard hasn't gone the notable extra step to invoke the centrally comic and seemingly illogical point that order can actually *arise* out of disorder. It is a point that would fully contradict the gloomy, dissipative second law of thermodynamics and the general cultural depression that attends such a dissipative worldview. In *Arcadia*, it "naturally" arrives from the least likely of all sources, Chloë, an innocent girl in the contemporary set of the play who is seemingly out of her intellectual realm. Virtually out of thin air, this untutored innocent makes perhaps the most telling observation of all: "The universe is deterministic all right, just like Newton said, I mean it's trying to be, but the only thing wrong is people fancying people who aren't supposed to be in that part of the plan" (73). Valentine puts it in more formal terms: "Ah, the attraction that Newton left out. All the way back to the apple in the garden" (74). What is *supposed* to happen and what *does* happen are clearly two different things. Life isn't random, but it *is* unpredictable. And human emotions are the "strange attractors" that make the world go round. "Declining to feeling" and accepting the chaos patterns that inform and even regulate our bodies (and therefore our lives) will allow us to appreciate "arcadia" as no traditional rational enterprise ever has.

While order in the traditional sense is thought-imposed control over the disorder of indulgent feeling, the comic and chaotics enterprise argues for a life model embracing order and disorder manifest through the interaction of thought and feeling. When one informs the other, the result is more likely to be a life well lived in scaled harmony with the "disorderly" order of all existence. We see the uncanny synonymity of chaos and comedy at work.

At this point, still, the all-important point that order arises out of and actually reverses disorder hasn't surfaced in the play so much as it's so far only been discussed. But it does find itself poignantly dramatized toward play's end. In rather coldly historical terms by the contemporary gathering of characters, we are first told that the world of Thomasina and Septimus was disrupted by a house fire that killed Thomasina in 1809 and left Septimus to take on a hermit's life trying to prove his protégé's theories. But then the play turns to Thomasina's last night on earth, when she and Septimus dance quietly together in a warm embrace that is in turn shadowed on stage by their twentieth-century counterparts, Hannah and Gus. The doubled dance is a waltz, and the two couples intermingle on stage in a final comment on the interrelationship of the two sets/worlds, which is a reflection in turn of the fractal self-similarity of all life as it thermodynamically whirls and stirs itself like rice pudding toward a heat death. But this dance involving one couple doomed to mortal separation and another couple making warm beginnings leaves the dramatic impression that life itself literally reverses the jam and pudding swirl of living into a counter-intuitive dance of self-organizing (re)union. Universal heat death is an inevitability, perhaps; but in the face of that inevitability occurs countless natural and human actions of significant, if impermanent, self-organization. Life, the highest form of self-organization, literally defies an otherwise pervasive universal movement toward utter, random disorder. Life regenerates, swims upstream, and prevails even against apparently overwhelmingly dissipative tendencies, arguing that humanity, finally, should be hope-full rather than despair-ridden.

The two dancing couples at play's end also circle us back to the play's opening, where Thomasina's very first question to Septimus reverberates throughout comedy: "What is carnal embrace?" (1). Septimus replies, coyly, that it "is the practice of throwing one's arms around a side of beef" (1). Life is certainly more than embracing a side of beef; but full possession of a good side of beef, like a good piece of reality, provides us with the sort of comic embrace of the world that argues with its own logic that life is far more than a random act in a world only *apparently* (tragically or absurdly) headed toward absolute zero. The comic impulse pulls itself out of any sort of despairing spiral into which humanity may find itself. This is why Erich Segal's announcement—and so many others' pronouncements as well—that comedy is dead is, from the logic of comedy, perhaps the most absurd statement of all.

Travesties Bounded in a Nutshell

In ages when dominant cultural paradigms perceived "lowly" nature as fundamentally hostile to "higher" human existence, the natural world almost invariably becomes the enemy to be defeated. From this perspective, malign nature actively pursues humanity in an effort to destroy the upstart promethean species; and for humanity to survive, subjugation of nature becomes the goal. The objective of human sciences in such times was to understand nature in a way that its revelations could be used to control that nature. And so the Rabelaisian gargantua known as natural impulse and rhythm was, from the comic perspective, erroneously and monomaniacally seen as a malignant consciousness bent on human destruction. The choice, deciding on a confrontationally tragic or a more embracing comic vision, belongs to humanity. The mind, then, that which constructs our visions of goodness and badness, virtue and vice, influenced by our vision of nature.

What we have is a complex network that acknowledges natural/material parameters within which cultural/material influences coalesce in the place called brain and named mind, creating an orderly/disorderly mix called "self."

Nature "naturally" inheres in brain, affected by the natural environment as well as the environment of culture to create mind and selfhood. What we have from the comic perspective is less a case of nature-seeking homeostatic order than a case of it seeking (to use Steven Rose's term) homeo*dynamic*[8] vitality. The equilibrium toward which consciousness and selfhood wends requires an unending dynamic interaction between and among our physical brain and physical surroundings. So claims the comic vision.

Daniel C. Dennett, in *Consciousness Explained*,[9] aligns with comic dynamism when he outlines a contemporary model of consciousness that argues consciousness is a continual process of instances of awareness along a "stream of consciousness" that never reaches a final destination but rather feeds back into itself at countless points along its neural progress. For Dennett, mind's contribution to knowledge and consciousness involves a stabilizing mechanism much like the stabilizing mechanisms of the eye. The eye jumps to many points of focus each second, but the brain levels off to a stable sense of panoramic vision. So, too, the mind manufactures a mental focus, drawing from myriad gathered data and creating an illusion of distinct, unified, and stable consciousness amid a mass of actually fragmented input.

In essence, according to Dennett, the mind takes a pandemonium of received data and an equally pandemonious list of possible reactions, and through a force of serialized manipulations considers the options and selects the best response for the moment, storing or eliminating that which is not included in the selected response for future consideration. Without this process of eliminating various data as "noise," human reaction would be impossible, faced as we are from moment to moment with a near infinity of material input and possible responses. The serializing operators developed by our brains as a crucial editorial survival mechanism allow for controllable human activity within a reasonable time frame and with minimized panic or confusion.

However, the process of eliminating what we habitually label "noise" restricts human ability for truly original change. New and truly beneficial frames of thought may be floating within the disorderly noise field, only to be ignored by serializing human habit. Finding the balance between serialized, causally deterministic mental control and unrestricted, unhindered, randomized material data input is the mind's function, the goal of art itself, and the message inherent in comedy. Too much order and we lose contact with the vitality of existence; too little order and we spiral out of control to a dissipative doom of oblivion. Too much and too little are both the domain generally occupied by tragedy. Finding a life between the extremes is the way of comedy.

In *Travesties* (1975), Stoppard took on the task of demonstrating that the artistic enterprise in general is the endeavor that best captures processes of human consciousness. From the mouth of the play's character James Joyce, Stoppard declares,

> An artist is the magician put among men to gratify—capriciously—their urge for immortality. The temples are built and brought down around him, continuously and contiguously, from Troy to the fields of Flanders. If there is any meaning in any of it, it is in what survives as art, yes even in the celebration of tyrants, yes even in the celebration of nonentities. What now of the Trojan Wars if it had been passed over by the artist's touch? Dust. A forgotten expedition prompted by Greek merchants looking for new markets. A minor redistribution of broken pots.[10]

The economic gist of Joyce's declamation indirectly refers to the Leninist position in the play, dryly presented in the opening monologue/lecture of act 2, wherein evidence of political artlessness is made clear by the discursive artlessness of the lecture itself. Stoppard dispenses with the artlessness of politics fairly quickly in this play, especially the artlessness that rigidly overlays a theory onto experience that simply doesn't fit the facts. Lenin, of course, changes history, but that change occurs rather tragically only by replacing one agelastic doctrine of capitalistic empire building with another encrusting theory of human sociality. Revolutions rarely do anything more than replace one system of artificial order with another. Unfortunately, it's the path humanity has chosen to follow all too often. This avenue of "change" doesn't seem to interest Stoppard, as it presumably would fail to interest any comic mind.

Joyce's reference to broken pots in the speech above is another matter, a theatrical reference to Tristan Tzara's Dadaist "art," which does seem to intrigue Stoppard. Tzara actually demonstrates his art-as-vandalism thesis in the stage actions that precede Joyce's speech: "*He starts to smash whatever crockery is to hand; which done, he strikes a satisfying pose*" (61). What is curious about the debate between Tzara and Joyce is that Tzara's advocacy of Dadaist randomness is less defeated by Joyce than by Tzara himself, who constantly rails against order/causality in one breath but then argues causally in the next. Arguing with the play's central character, Henry Carr, Tzara remarks, "You ended up in the trenches because on the 28th of June 1900 the heir to the throne of Austro-Hungary married beneath him and found that the wife he loved was never allowed to sit next to him on royal occasions; except! When he was acting in his military capacity as Inspector General of the Austro-Hungarian army" (40). Causal argumentation should come from anyone but Tzara. But even the anti-logic of Dadaist self-destruction has built logical roots. And implicitly, the destructive vitality that is Dadaism suggests that from such destruction will arise some sort of new order. It's just the case that Tzara's mission involves only the first part of this agenda, namely tearing down the facades. It deconstructs orthodox values without providing any follow-up ordering blueprint for reconstruction.

Joyce's argument, on the other hand, does include the element of disorder that Tzara seems exclusively to prefer. In Joyce's picture, without the catastrophes of the Trojan War or the Flanders Field of World War I, there would be no need for art and its ordering impulse. The significant difference between the two antagonists' arguments is that Joyce's argument additionally embraces the orderly disorder continuum as a unit, whereas Tzara is engaged in only the disorderly extreme of the continuum. The result is that what Tzara calls art—deconstruction and random expression—is an incompletion, recalling Stoppard's famous 1973 observation in *Artist Descending a Staircase*: "Skill without imagination is craftsmanship. . . . Imagination without skill gives us modern art."[11] Whichever applies to Tzara, his is not the artist's vision Stoppard seems to endorse. Rather, Stoppard's vision seems more in line with Joyce, an inclination confirmed by Stoppard himself in numerous interviews confessing as much. But it is only an inclination.

Between the two lies Henry Carr's rare vision, which is in fact the play proper, Carr's dramatized recollections of his life's experiences with these giants of history. Carr's suspect memory casts a serious shadow over the play regarding matters of historicity, factuality, and "truth." "Time slips" abound throughout the play, recasting single events as many as five times. Carr's pandemonious brain is at work here without the stifling influence of a causally obsessed, singular mental linearity. These time slips, however, eventually circle back around to a singular perspective that highlights the point that Carr is molding his entire mental output into a singular life history that has been informed by the one major imprint on his brain—his brief moment in life as an actor in the amateur production of *The Importance of Being Earnest*. Love interests, events of various sorts, speeches, virtually everything in Carr's consciousness finds its filter through his recollection of the amateur production.

It appears that Stoppard is suggesting that multiple drafts are our best chances at arriving at truth—that singular results can be of as dubious value as Carr's *Earnest*-informed visions of reality. Memory's single-draft inclination is humanity's singular encrusting flaw. In other words, the more fully we grasp and absorb the pandemonium of life around us and the more fully we allow our brains/minds to interact with those myriad experiences, then the more fully we live a life comically sustained. Pathetic in this world of fortunate Swiss neutrality, Carr's idealized encrustation escapes a tragic doom by simply having been placed in a hurricane's calm center surrounded by turmoil that has been generated by the very flaws of perception and logic he exhibits. He's a fortunate fool surrounded by a less fortunate world bent on self-destruction because of the same fundamental thought processes that Carr exhibits. Inflexibly linear, grand European culture finds itself imploding under the logical weight of its own dissipative arrogance.

But in the end, it's not any of the play's characters that really stand out as full comic visionaries. That honor goes to Stoppard alone whose position is encapsulated in the play as a whole. Absorbing a greater mass of pandemonious material and finding a way to organize the material diversity he absorbs, Stoppard's mind bounds well beyond editing factuality into a linear line of thinking. He infuses that wealth of input with far-reaching, imaginative mental interplay that expands rather than constricts the vital operations of the mind upon the material his senses have gathered for him to consider. He has allowed his mind to play with and to create an integrated reality that is aware of its truth-dealing limits but little caring to be chastened by an idealized awareness of those limits. There may never be one truth from this point of view, but such a truth really doesn't matter. The play's the thing wherein to catch the quality of our consciousnesses and therefore of our lives. Trying to secure mere factuality has its actuarial place in the world, but comic reality revels in experimenting at warping, stretching, and coloring its way to truth, creating in turn levels that place us in the world we're otherwise observing and trying to mold from afar. And this point of play is important because when the mind follows the body and accepts its role as part of that world, then it can tap into comic exuberance and is one step closer to finding its place in an order that can improve rather than hinder the survivability of that body in which it is housed.

Jumpers: To the Rescue

In his 1972 play, *Jumpers*, Stoppard utilized his pandemonious comic sensibility to create his own chaotic "Thinkery," a world of sophistic argumentation most likely in this case between two *wrong*-thinking schools of thought. In particular, Archie and the recently deceased McFee represent a school of acrobatically agile thinkers allied to pragmatic reductionism and moral relativism, arguing generally for a self-preservationist philosophy of life. George is the sympathetic dunderer of the work, following philosophical strands of thought in order to derive a proof that moral absolutes exist and that they are based on our responsibilities to others.

As George rehearses his lecture on the proof of god's existence via the "First Cause" argument, he confronts Zeno's famous paradox of infinite regressions, which basically argues against common sense and casts doubt on any human ability to derive unassailable conclusions about virtually anything. George argues that the notion of infinite series "[l]ed the Greek philosopher Zeno to conclude that since an arrow shot towards a target first had to cover half the distance, and then half the remainder, and then half the remainder after that, and so on *ad infinitum*, the result was, as I will now demonstrate, that though an arrow is always approaching its target, it never quite gets there, and Saint Sebastian died of fright."[12] By this reasoning, one must accept Cantor's Proof that the arrow could not move at all, in fact that all motion is illusion. But George empirically refutes this obviously absurd though logically demonstrated assertion by letting an actual arrow fly. The physical world contradicts the best of abstract logic, defeating the complex sophistry of Zeno's paradox. George's common sense seems to prevail.

But in more significant ways, George is anything but a defender of common sense virtues. As philosophical dunderer, he is part of that long tradition of men too intellectual for their own good. For example, in a palpable reversal of intellectual fortune, even as George is arguing for such moral absolutes as "Thou shalt not kill" and building his argument against relativism, the material world crashes hard into his musings. As George practices the arrow demonstration to refute Zeno, his arrow strikes his pet rabbit, killing it and thereby rattling his ideas regarding the sanctity of life. To add further injury to his exalted thinking, literally staggered by the realization that he killed his rabbit, he additionally inadvertently steps on and kills his pet tortoise. It seems that rather than arguing for the sanctity of life as a noble idea, George should actually practice what he preaches and pay more attention to the living world around him. Actions trump ideas at the very least in the cases of George's two pets.

From this humorous episode, Stoppard moves to a far more serious but related matter. George's obsession to know the world and its rules by locking himself away in his study and pursuing intellectual threads of thought effectively leads him to commit the one genuinely "unpardonable sin" of the comic agenda. His wife, Dorothy/Dottie, is going mad as a result of taking to heart the news item of the day that two British astronauts are marooned on the moon and fighting to the death for the single opportunity to return to earth on their crippled space ship. One astronaut must remain behind. This background event in the play is an apparent proof that life is indeed nasty and brutish, that self-preservation prevails over the virtue of self-sacrifice in the world. Dottie/Dorothy, a dreamy-eyed showgirl whose musical staples all involve romantic ditties glorifying the moon, can't accept this apparent truth about the world and clearly needs George's loving attention to overcome her crisis of faith in humanity (and the moon). But George critically misses every opportunity concretely to "prove" his humanity and demonstrate the selfless humanistic compassion his philosophy sets up as the highest human ideal. George remains caught up in his study dryly trying abstractly to *prove* a humanistic perspective instead of physically practicing his humanity through simple action. All would be won if George simply chose to care for his fellow human being—and a spouse at that—rather than to stow away in his Thinkery. His inattention is not altogether different, in fact, from the selfish behavior of the astronauts or from the selfish endorsements of his philosophical opposition. As Bergson would warn, failing to fulfill one's social duties is the root of all human suffering. Individualized self-absorption destroys the human fabric. Even while striving to represent the side of selfless virtue, George remains oblivious to the obligations he fails to fulfill.

Yet another material inconvenience throughout the play threatens to impose itself on George's reflections: the murder (or possible suicide) of his colleague and intellectual rival, McFee. Following McFee's death, Archie takes charge of the survivalist argument by in fact becoming the personification of the position by having risen to prominence as a result of a companion's death. It is a death that confirms Archie's (and McFee's) amoral, relativist position that the death of others is frequently a good thing and perhaps should be encouraged especially in situations where it improves one's chances of extended or enhanced survival. How can anyone prove otherwise? If improved chances of personal survivability are at stake, is it not correct and ultimately justifiable to do what is necessary to improves one's chances? If stranded on the moon, shouldn't one do anything necessary to survive?

George fights this general position but ultimately fails to construct an argument to prove his position. Even Dottie is temporarily won over to Archie's side. Feeling badly about a whole string of things, Dottie reports to George how Archie has tried to console her: "[S]ometimes he makes them seem not so bad after all—no that's wrong, too; he knows not 'seems.' Things do not *seem*, on the one hand, they *are*; and on the other hand, bad is not what they can be. . . . Things and actions, you understand can have any number of real and verifiable properties. But good and bad, better and worse, these are not real properties of things, they are just expressions of our feelings about them" (27). Dottie's summary of Archie's/McFee's relativist position reminds us of the Hamlet-esque point that mind can effectively argue anything to be true if given free reign. But rather ironically, innocent Dottie—like Chloë in *Arcadia*—hits on a key point. Archie has moved good and bad from the list of qualities inhering in given actions, arguing that they are mere descriptors of "our feelings about them." However, Dottie's Hamlet-esque speech speaks volumes to why she is present in the play, anticipating what Damasio articulates about feeling: "Consciousness and emotion are *not* separable."[13] Feeling provides the basis for thought and by extension is—or should be—the cornerstone of reason itself. George and Archie each "thinks" in emotional isolation, anxiously or smugly drawing "unfeeling" conclusions of the world by sidestepping the verities that invade us daily through our habitation in our bodies. Dottie, on the other hand, *feels*. And it is the feeling of wrong in the end that actually *is* rather than *seems* the correct response.

Archie indulges the world of carnal embraces for purely selfish reasons, a "perfect survivalist." Almost a perfect villain, complete with charm and social graces, Archie can shrewdly manipulate information to serve his selfish desires. As Dottie's speech implies, in this present sort of world, Machiavellian seeming dominates being, though in a quick-witted and humorous Stoppardian manner. Dottie, however, knows and does not seem because she accepts her feelings as viable originary points from which to develop understanding. Feelings, intuition, and a mind attuned to them in harmony with the world around her are what make Dottie human and is finally why the astronauts' behavior so horrifies her. That lack of human fellow feeling makes Archie and even George something less than human, their heightened intellectual capabilities notwithstanding. George and Archie are disconnected from their feeling and therefore from the world, in violent contrast to Dottie's connectedness.

In the play's closing coda, George makes a telling observation: "There are many things I know which are not verifiable but nobody can tell me I don't know them" (69). Reason, it seems, even for George may finally not be the only order of the day. At this point, George is momentarily slipping into Dottie's world, which finally trumps both George's and Archie's worlds as worlds worth living in. Comedy is very wary of arrogant but empty reason, always preferring feeling when the choice is between it and reason, but fully endorsing the more rounded occasions when feeling weds reason. Sloppy at times, illogical and unintellectual, its unifying vitality provides opportunity to break through Zeno-like logjams of thought.

Moving beyond the folly of George and Archie, Stoppard inserts a Dottie-like point into the play that argues her case better than any philosophical treatise. The catastrophic mission to the moon seems to have empirically "proven" the accuracy of Archie's reductionist, amoral vision of the universe with almost Petri-dish scientific precision. Seeing the events unfold on the lunar landscape, we are presumably given with crystal-clear accuracy an irrefutable proof that self-interest will prevail over moralistic concepts like self-sacrifice. However, against this gloomy evidence of man's "baser" natural drives, Stoppard stirs a counterexample into the mix: the recollection of an Antarctic explorer "sacrificing his life to give his companions a slim chance of survival" (70). An apparent case of true altruism leads McFee himself to conclude that "[i]f altruism is a possibility," then his (and Archie's) self-serving philosophy "is up a gum-tree" (70). The Antarctic explorer's altruism is put forth here as physical evidence of a curious human disposition toward a paradigm of cooperation and giving that blasts Archie's (and McFee's) selfish, Hobbesian, and entirely logical conclusions about humanity's innate selfishness. It seems entirely logically wrong-headed for humans to engage in such altruism. If altruism is possible, in fact, reason itself is in jeopardy.

If the first principle of life is survivability, then altruism must surely be the height of illogical, counterintuitive behavior. But if somehow humanity is wired to behave altruistically, then logic itself comes under fire. So is altruism somehow a survivalist virtue and the logic of self-preservation flawed, as George vainly tries to argue? Or is altruism an illusion and the logic of survival our first principle, as Archie argues? Can, impossibly, the answer to both be *yes*? Comedy seems to think so.

How Altruism?

In comedy, much rides on the potential to engage logically disruptive—and beneficially disorderly—altruism among humans, that category of behavior that willingly relinquishes something in one's possession for the betterment of an other. This is the basic cornerstone of virtually all communal agendas, the heart of the mantra "from each according to ability, to each according to need." It is also perhaps the only possible way to disrupt pathologies of oppression without merely replacing one oppressive order with another. But, given that this fundamentally comic dream has been placed before humanity now for millennia, is it little more than an impossible dream never to be realized?

Realizing that nature is a dynamic system that thrives on change brought on by an interplay between order and disorder, evolutionary theory nonetheless recognizes that nature generally aims for stability and even stasis. Recall the second law of thermodynamics and the entropic paradigm. Even among living organisms, there exists what John Maynard Smith has labeled evolutionarily stable strategies.[14] Scientists will also be quick to remind us that self-reflective terms like "thriving," "recognizing," and "aiming" have no place in descriptions of natural behavior because nature lacks a consciousness to be aware of such behavior. And herein lies part of the problem of the comic agenda as generally perceived. The generally logical insistence that humans (or any living organism) would be better off cooperating rather than fighting and oppressing each other requires a consciousness that rises above natural predicates that operate beneath consciousness and that function purely to ensure individuated survival. In fact, evolutionary science prior to the 1960s may have misled at least a few recent comic theorists, given that, as Robert Axelrod observes, most ideas regarding the Darwinian theory of survival of the fittest stem from a "misreading of theory that assigned most adaptation to selection at the level of populations or whole species."[15] What has become increasingly evident, however, is that evolution functions at levels far more individuated than speciation, even to the point that Richard Dawkins, in his landmark *The Selfish Gene* (1976),[16] convincingly argues that survival strategies originate at the level of genes and that the things we call discrete bodies—throughout nature—are nothing more than survival machines designed by genes to promote the perpetuity of their existences by way of self-replication and transferal of myriad duplicates "selves," the success of which depends on a given gene's greatest environmental suitability. In other words, consciousness of what "ought" to occur at, say, the species level has no real place in this world where literal sensitivity to what must immediately occur for survival exists purely at the individualistic level.

But the level to which one applies evolutionary processes (down to genes or perhaps even lower [or even slightly higher]) is less important here than it is to note the general point that, in nature, there is no awareness of nor inclination toward seeing anything in terms of species or groups. And so, speciated explanations accounting for apparent group behavior, any conclusion that certain behaviors in nature are designed for the good of the species, really are misreadings of individuated actions among incredibly selfish units of existence that are merely generating a comfort zone that will provide for the greatest possibility of individuated replication and transfer of identifying features (genes, DNA, etc.) into perpetuity. Individuated self-interest rules the day, even in instances of apparent cooperation because they invariably involve little more than actions undertaken to increase chances of individual survival. Every living organism, down to the basic building blocks of life, is obsessed with personal survival, frankly uninterested in assisting anyone or anything else up the ladder of success. The genetics perspective that Dawkins advocates, for example, explains moments of apparent altruism among genetically related individuals. Sacrificing oneself for the good of one's offspring may appear to be altruistic, but it actually is no more than our selfish genes at work engaged in protecting their replicated selves as they move into the next generation of selfish competition. Altruism, in short, is a natural illusion.

At best, altruism is a term that rather inaccurately describes some level of reciprocally beneficial cooperation. Reciprocally beneficial cooperation *is* something found in nature, and that is at least an important first step. The herding instinct of certain organisms, for example, while *looking* like it is designed to preserve the species, is really nothing more than a band of individuated creatures utilizing a mutually beneficial tactic that ensures a higher degree of survivability than would exist had such creatures decided to go it alone. In fact, evolution explains why there is no such thing as a nonherding gazelle: if such creatures ever existed, they have long since become prey to their carnivorous neighbors, leaving the field to their more successful herding counterparts.

So one option for humans in their utopian and counter-natural efforts to fulfill the comic vision of interpersonal cooperation and of abandoning our pathology of competitive oppression is somehow to *argue* that we as a confederation of individuated beings would actually each individually benefit by engaging in cooperative enterprises that only *appear* to be altruistic but are actually interactively beneficial to *all* parties involved. But this argument must convince even and especially those individuals who somehow feel they are better served by continuing the individuated practice of standing alone and oppressing others. After all, we have rather overwhelming evidence that simply arguing from some moralistic/idealistic high ground that altruism would be for the greater good of the species by itself won't work. In its naïve manifestations, it's what comedy does, and it gives comedy a bad name in the process.

While such an argument *may* convince some among us, it needs to reach that point where group generated profitability overwhelmingly outweighs the counter-urge to pursue individual profit. Even if something like that does happen, however, the best we've achieved is mutual beneficiality, a cloaked version of self-interest likely to fall apart at the first opportunity for personal gain. Actual sacrifice for the good of the group is not part of this equation, for even if idealistic/rational/moral arguments could somehow convince even a majority of humanity that helping others to the point of personal sacrifice is what humanity should do, at least some among us would take advantage of a system that endorses sacrifice. In an environment where sacrifice is encouraged, duplicitous survivalists will truly thrive.

In *The Selfish Gene*, Dawkins draws from a wealth of game theory insights to make just this point. He demonstrates the point by creating a hypothetical community in which a vast majority (not just a paltry few) of its individuals is willing to help each other when a need arises. Dawkins hypothesizes a species that has a need to have fellow members remove potentially deadly parasites from unreachable parts of each other's bodies. Into this group, however, he inserts members who accept the "altruistic" service of others but give no assistance in return. The former he calls suckers, the latter, cheaters. Evolution will reward the cheaters, who benefit from others' expenditure of energy without themselves expending energy in like service for their fellows' benefits. The result is that the suckers will eventually fall into extinction, having exhausted their resources and increasingly received no benefits in return. The cheaters will prevail.

However, following the example to its inevitable end, we also see that while the cheaters will dominate the community, their selfishness will ultimately doom the species to extinction, given that the cheaters will eventually be at the mercy of a parasitically overrun environment. Selfishness has destroyed the species, with the altruists succumbing first. Reason *may* prevail in such circumstances and *may* convince individuals to think about the good of the species with the result being that the species *may* in fact thrive in this new order. This is what naïve comedy tends to hope will occur. But nature suggests otherwise.

Into this mix, Dawkins offers a hopeful possibility by inserting a third type, what he calls the grudgers. They assist everyone, suckers and cheaters alike, but once they reveal a cheater by verifying its refusal to reciprocate, the grudger develops a memory that will exclude assisting that cheater in the future. Dawkins reports the results of a computer simulation of the interactions among these types:

> The first thing that happens is a dramatic crash in the population of suckers as the cheaters ruthlessly exploit them. The cheats enjoy a soaring population explosion, reaching their peak just as the suckers perish. But the cheats will have the grudgers to reckon with. During the precipitous decline of the suckers, the grudgers have been slowly decreasing in numbers, taking a battering from the prospering cheats, but just managing to hold their own. After the last sucker has gone and the cheats can no longer get away with selfish exploitation so easily, the grudgers slowly begin to increase at the cheats' expense. Steadily their population rise gathers momentum. It accelerates steeply, the cheat population crashes to near extinction, then levels out as they enjoy privileges of rarity and the comparative freedom from grudges which this brings. (186)

The suckers are indiscriminate altruists, a noble but endangered breed whenever they exist in a world occupied by any but fellow suckers. Even one cheater can eventually destabilize this utopian arrangement. And in a world that includes the more self-serving but still benevolent grudgers, the suckers remain dangerous not only to themselves but also to the grudgers: "[T]he presence of the suckers actually endangered the grudgers early on in the story because they were responsible for the temporary prosperity of the cheats" (186).

Anyone who has considered the vulnerability of generosity among humans has likely hit on some of the same conclusions drawn in Dawkins' game: pure altruism is a sucker's behavior, vulnerable to cheater's abuse; and even a reciprocal altruist (the grudger) with more tempered actions is vulnerable to abuse and even prolonged periods of endangerment, even the threat of extinction. This scenario is a rather simple model and comes nowhere near to capturing the nearly overwhelming complexity of human existence. But Dawkins' model does effectively highlight the dangers inherent in any unilateral attempts to change the world by altruistic means. This is where a more hard-nosed brand of comedy must find its way. A utopian dream crying out that we must all just somehow get along simply won't work.

A Silly Little Thing Called Love

Kelly Oliver concludes her 2001 study on the culture of oppression, *Witnessing: Beyond Recognition*, with the following observation: "Relations with others do not have to be hostile alien encounters. Instead, they can be loving adventures, the advent of something new. . . . In the thrilling adventure of love, the unknown and incomprehensible excite rather than threaten. Falling in love . . . is the greatest joy; and vulnerability in the face of the other is a sweet surrender, a gift rather than a sacrifice. The other's potential to make me better than I am is the power of love. . . . And being together is the chaotic adventure of subjectivity."[17] Oliver captures here the comic spirit at its most romantic, but also at both its most critical and vulnerable points. Exactly how does one slip into this brave new world of loving gift without falling victim to the cheaters who prey on such giving vulnerability? Love is indeed a force of nature to be reckoned with, as Chloë and Dottie—and so many others—have pointed out. But accepting and indulging love, as we all surely know, exposes our vulnerable selves to rapacious others by placing us squarely on the pedestal of suckerdom.

By all accounts, Oliver seems to be speaking of a love that dominant culture would at the very least be able to allow within certain constraints to exist, generating as it does an impulse to people the world and therefore to prolong the selfish life of the endorsing culture. But what of the sort of love that doesn't directly or immediately appear capable of serving culture's selfish end? In *The Invention of Love* (1997), Stoppard brings this very question into focus by dramatizing the life of A. E. Houseman, a homosexual man whose celibate life of regret was in large measure the result of hiding from the virulent dominant-culture homophobia of his native Great Britain. Stoppard contrasts Houseman's life with the far more notorious life lived by his contemporary, Oscar Wilde. In both cases, agelastic constraints on genuine though unorthodox love crush the full, vital potential of these two men. By extension they exhibit the lost potential of all living creatures forced onto the procrustean bed of social order.

Stoppard, however, rather ingeniously draws on an actual (or at least mythical) precedent when society at least once in history opened itself uncritically to the full power of this genuinely awesome force of nature. Housman reports a Classical Greek account of the creation of "[a]n army, a hundred and fifty pairs of lovers, the Sacred Band of Theban youths, and they were never beaten till Greek liberty died for good at the battle of Chaeronea. At the end of that day, says Plutarch, the victorious Philip of Macedon went forth to view the slain, and when he came to the place where the entire three hundred fought and lay dead together, he wondered, and understanding that it was the band of lovers, he shed tears and said, whoever suspects baseness in anything these men did, let him perish."[18] Clearly, love can be a transforming force possessed of and generated by a logic of its own, capable of inspiring sacrifice and a unifying strength far greater than the sum of its parts. But can it be of any use, beyond the brief moments of joy it brings to couples huddled beyond the stern eye of moral censors? When allowed its way, Stoppard suggests, love is capable of changing the world. But how does it find its way?

CHAPTER 6

Comedy as Gift

Trying to solve the mystery of McFee's murder (perhaps suicide) in *Jumpers* and describing McFee as his intellectual mentor, inspector Crouch observes that McFee "kept harking back to the first Captain Oates, out there in the Antarctic wastes, sacrificing his life to give his companions a slim chance of survival. . . . Henry, he said, what made him do it?—out of the tent and into the jaws of the blizzard." It's at this point that Crouch reports, "If altruism is a possibility, he said, my argument is up a gum-tree."[1] McFee is at very least logically correct for backing an anti-altruistic platform. After all, such altruism is an illogical and counterintuitive proposition given that Darwinian self-preservationist behavior is the law of nature. It is an irrational behavior from a cultural-dominant position as well, given that it disrupts an orderly system of investment and return (economic or otherwise), which is the cornerstone of virtually every Western orthodox system of human interaction.

Consider that society is fundamentally based on the concept of maintaining equilibrium by enforcing a system of exchange that maintains the life of the whole group. Giving without expectation of return just doesn't make sense on a "civilized" social front or even from a typically conceived natural perspective. How can order be sustained if people don't follow proper rules of exchange? Spontaneous eruptions of altruism could actually call into question the validity of "orderly exchange" as we now know it and invite a reevaluation of the institutions built up to promote and defend such systems, which would of course be unwelcomed by those in control. If altruism somehow were to take hold, it would be every bit as threatening to society—and those in power—as an assault by steel and gunpowder, perhaps even more so.

Of course, society in this case is simply following the lead of its component units, the citizens themselves. When it comes to goods and services, it is quite "natural" to want, at very least, an equal return on investment. And frequently we go in the other direction, looking for greater returns on—actual profits from—our investments. It's a point that clearly falls in line with our apparent instincts to improve our survival odds, even (or especially) at the expense of others. And by extension, we surely never ask for less than we invest. In fact, most systems solidify their positions on the matter by arguing that the idea of a *true* gift is a mere delusion because at very least the gifter himself or herself would have an unfulfilled sense that reciprocity is in order, despite not being specifically requested. Could it be, for example, that Captain Oates' sacrifice had buried in it a sense that his heroism would reap appreciative benefits that would eventually go to his surviving family members back home? Or would the immortality of a position among his nation's honored heroes be reciprocation enough? It really is difficult think of any action as not entailing at least the expectation of some element of reciprocity.

Jacques Derrida enters the matter on the side of McFee by arguing that a "true gift," freely given without any design on return and without any pressure on the part of the receiver to feel indebted, is a *logical* impossibility. According to Derrida, for true gift giving to occur, the giver must remain anonymous in order not to receive so much as a reward of thanks, and the receiver must actually not realize he or she is a beneficiary in order not to feel indebted. In fact, the giver must actually forget he or she has given a gift in order not to experience the reward of personal satisfaction. Anything short of giver and receiver oblivion would beg a consciousness of indebtedness or gratitude, either of which is a sort of reciprocation that restabilizes the system by denying the imbalance generated by the presence of a pure, unreciprocated gift. Gifting becomes a logical impossibility within such a system, and believing in pure gift, believing in true altruism is, in a word, *madness*.

This impossibility, however, presumes that current systems of logic and reason inexorably define the world. Something like comedy, unfettered by mere logic, provides an alternative to the closed, agelastic systems of order governing human existence. Comedy and its giving/altruistic agenda argues a destabilizing possibility that explodes current social economy. In many ways it explodes any attempt at philosophical rigor whatsoever, revealing the impossibility of encapsulating human action in a closed system. Gift disrupts the circle/cycle of reciprocal giving and receiving in ways that reason and philosophy seem incapable of accepting.

This point recalls yet another of Richard Dawkins' insights in *The Selfish Gene*. While he spends a great deal of time dealing with the reciprocating behavior found in nature, Dawkins also ventures into the realm of human culture, suggesting human culture too is a soup of ever-changing opportunity much like nature is. Out of that human cultural realm has evolved what he calls "memes," described as "a unit of cultural transmission."[2] Memes operate within human consciousness like genes, with the "good" ones replicating in human minds and getting passed on. But unlike the long process of genetic evolution, these mentally generated survival units actually develop and "propagate" at a faster rate than biologically generated genes because these ideas, thoughts, and actions that increase our chances of survival are passed on to others through education and other means of socialization. And those that are detrimental or inefficient die out. From this perspective, it is therefore at least *possible* to conceive that after millennia of comic percolation, humanity could slip into a collective consciousness that sees survivability at a species-conscious level and develop an altruistic meme that could at long last overrun our genetic tendencies toward selfishness. It's a tantalizing possibility. But much has to be accomplished within our sense of sociality before such a transformation can occur.

Consider these points in light of Derrida's analysis of the gift. He sees human sociality as being fundamentally a closed circle of interactions and then observes,

> The overrunning of the circle by the gift, if there is any, does not lead to a simple, ineffable exteriority that would be transcendent and without relation. It is this exteriority that sets the circle going, it is this exteriority that puts the economy in motion. . . . If one must *render an account* (to science, to reason, to philosophy, to the economy of meaning) of the circle effects in which a gift gets annulled, this account-rendering requires that one take into account that which, while not simply belonging to the circle, engages in it and sets off its motion. What is the gift as the first mover of the circle? And how does it contract itself into a circular contract? And from what place? Since when? From whom?[3]

This rather protracted passage asserts that the gift must existentially originate beyond any standard circle of exchange in order for a circle of exchange *to be* in the first place. Despite the logical impossibility of locating a first mover (recalling Zeno), a first mover as gift giver must exist somewhere prior to and as originator of the order we now experience. Nature in all its bounty is itself a gift that that is unearned by its current beneficiaries. Here we see the idea of god (first mover) as gift giver (of life, etc.), the "entity" (whatever it might be) that begins the circle as it stands outside or beyond the system it has generated. Though not initially belonging to the system, gift and gift giver are necessary for the system to be. And in our current post-genesis world, theoretically at least, whenever it is that a similar (though logically impossible) gift from the outside affects the circle, the gift eventually becomes absorbed into the circle it has influenced/generated, initiating a new or renewed or revitalized circle, perhaps more just or more equitable in its new manifestation.

The idea that gift giving is the first cause of new orders has a certain almost titanic appeal, following as it does the singular "First Cause of Creation" in that it has similar life-generating and life-altering characteristics and has a certain appeal. For example, an existent system in decay may require a gift of rejuvenation while a closed, agelastic system of strict reciprocation forbidding gifts of forgiveness or grace can enslave all involved in an unending cycle of retribution. Recall that Hamlet observes, "Use every man after his desert, and who shall scape whipping?" (II.ii.516–17), addressing the postlapsarian point that human nature is lost if it is required reciprocally to pay for or earn its own salvation. In such situations, a gift of rejuvenation that cannot be repaid by the recipient seems humanity's only hope. The closed system of precise exchange *may* have its place in the commercial marketplace, but its more universal application to interpersonal, moral, or metaphysical matters seems inappropriate at best and more often catastrophic at worst.

If humanity must only reap what it sows, catastrophe will inexorably ensue. In theological terms, the gift of grace is humanity's saving mechanism, a cosmic apparatus of gratuitous benevolence based precisely on the notion that the recipient doesn't deserve the rewards of life and happiness but is given such rewards nonetheless. And the recipient is never in a position fully to reciprocate since the full gift of grace is invariably reserved for an omnipotent dispenser—God, or force of nature, or Comedy personified. But it is sometimes also reserved for a near-omnipotent monarchic *deus ex machina* (recall Molière's *Tartuffe*), grand though perhaps only wishful examples of an essentially benign and even actively benevolent governing *human* force. While we are not currently memically programmed to see such ideas as anything but utopianly wistful thoughts, is it not possible, perhaps, to continue to pursue a memic transformation of thought that welcomes genuine altruism without a sense of debt, which in turn would rejuvenate the natural and cultural circles within which we live?

One concrete way of memically visioning the above as possible is to perceive existence in Rabelaisian terms, though updated a bit. Our culture generally perceives human interaction as operating in a "zero-sum" fashion. That is, for anyone to advance, it must be at the expense of someone else, a presumption of course that leads to a balancing act that ultimately develops into a pathology of oppression. However, it is possible to see at least some of our fundamental interactions as non-zero-sum enterprises when advancement need not occur at the expense of others. That, in short, is the Rabelaisian vision, taking in the bounty of nature as a gift that benefits all to the detriment of none. Challenging the rationalist argument that denies the concept of the non-zero-sum ethos of gifting plays a crucially important part in the comic agenda. Furthermore, the point that comedy involves itself so concretely with social exchange suggests that comedy and yet another branch of human understanding are not so distant relatives. Economics plays its part in the grand comic scheme as well.

Gift Economy and Exchange Economy

Marcel Mauss' landmark 1925 study, *The Gift: The Forms and Reasons for Exchange in Archaic Societies*,[4] introduced the idea of the gift to contemporary analysis in various disciplines. Curiously, though, as Derrida notes, "Marcel Mauss' *The Gift* speaks of everything but gift" (24) in that the gift economy Mauss analyzes is exactly that, an *economy*, which correctly Derrida observes "implies the idea of exchange, of circulation, of return" (6). For there to be a "true" gift, on the other hand (again, quoting Derrida), "there must be no reciprocity, return, exchange, countergift, or debt" (12). Mauss' concept of gift exchange differs radically from the one Derrida describes (and which Derrida concludes is impossible), focusing on a basic social bond derived from gift giving where return is actually expected.

Mauss' conclusions regarding gift identify a long list of interrelationships that originate from gifting, running from benign social bonding to the notorious "potlatch" wherein the giver aggressively gives until the receiver is no longer able to return in kind, thereby marking the giver's dominion over the receiver. Mauss sees that these archaic exchange patterns each have a sort of value-added attachment in that the gift itself necessarily contains a *trace* of the gift giver, a point not altogether different from Derrida's. It is actually a case of the giver giving something of his or her *self* to the other. It differs from Derrida, however, in that such intimate giving is self-consciously intended to create a bond of a kind barely remembered in strict commodity exchange economies such as Western culture has adopted. According to Mauss, "giving [a gift] is giving *oneself*, and if one gives *oneself* it is because one 'owes' oneself—one's person and one's goods—to others" (46). Mark Osteen explains that "[t]he persons who live in these societies, in other words, represent themselves not as the self-interested individuals of neoclassical economics but as a nexus of social obligations."[5]

Mauss' conception of the gift as social bond reveals the weaknesses of the rationalist individualism that permeates contemporary Western culture. The Western concern over whether or not "pure" gift (as described by Derrida) is even a possibility involves the idea of an autonomous individual sacrificing economic gain for the economic benefit of another autonomous individual. What is fundamentally missing in this concept is the selfless element of bonding that gifting in a gift economy generates. The Western market economy of equivalent exchange measures capital value freed from what the West would call nostalgically unmarketable sentimental value. Such sentimental gift value has generally been stripped away in Western market economy and yields to practical matters of useful value alone.

What Mauss suggests as necessary for a return to more viable and integrated social units is to reunite "concepts of law and economics that it [currently] pleases us to contrast: liberty and obligation; liberality, generosity, and luxury, as against savings, interest, and utility" (73). In his incisive analysis of gifting, Andrew Cowell sees this manifestation of the gift as a potential "disruptor of the cycle of the West, the masculine West, the capitalist West, and the logocentric West."[6] Much like the metaphysical concept of gift described earlier, this socio-anthropological idea of the gift likewise disrupts the economic cycle of utilitarian exchange when it enters into the circle from its originary outsider's position, and in the process it unites individuals in a social compact beyond mere utilitarian-valued reciprocity.

Lewis Hyde in *The Gift: Imagination and the Erotic Life of Property* (1979) clarifies the distinction between Western economics and the economic model that relies on gift exchange by describing the history behind the phrase "Indian giver." Western settlers of the New World were confused by the Native-American practice of giving and then expecting a gift in return as well as the practice of passing on gifts received from other givers. Hyde observes of the clash of exchange philosophies: "The opposite of 'Indian giver' would be something like 'white man keeper' (or maybe 'capitalist'), that is, a person whose instinct is to remove property from circulation, to put it in a warehouse or museum (or, more to the point for capitalism, to lay it aside to be used for production)."[7] Hyde emphasizes that in this system, "[a] man may wonder what will come in return for his gift, but he is not supposed to bring it up. Gift exchange is not a form of barter" (15). And a second point, drawn from Bronislaw Malinowski's study of South Seas gift economies, "'is that the equivalence of the counter-gift is left to the giver, and it cannot be enforced by any kind of coercion.' If a man gives a second-rate necklace in return for a fine set of armshells, people may talk, but there is nothing anyone can do about it. . . . You put yourself in his hands" (15).

This idea of gift is by no means as "pure" as Derrida seeks, but it reveals a transformative design nonetheless, unencumbered by tight logical constructs. Free will is in effect; decisions concerning return lack strictly reciprocal obligations of material equivalence while acknowledging social bonds without regard to "intrinsic" worth. Hyde notes that "market exchange has an equilibrium or stasis: you pay to balance the scale. But when you give a gift there is momentum, and the weight shifts from body to body" (9). If gifting seeps into the utilitarian, static exchange system of equal reciprocity, a risk of disequilibrium arises as does the more dangerous possibility of the eventual implosion of order. Clearly, gifts are dangerous things in a market economy and for a society founded on such an economy. The very stability of the autonomous individual is in jeopardy.

What Mauss has discovered, however, is that gift giving and receiving—especially when the gift has high sentimental value—involve *aneconomic* expansion of oneself by giving and receiving simultaneously without feeling any need whatsoever for equivalent return. In other words, giving and getting inserts us into our *place* in the world not as autonomous individuals isolatedly accumulating wealth, but as citizens of a living social organism. Through gifts, we break apart the closed circle and enter into it by adding to it. And the result is an improvement on the circle we've entered. If such a system were fully implemented, participants would become "enworlded" into a tangled, inextricable mass that would undermine the neoclassical conceptions of individualism, which is the foundation of Western commodity exchange. Value would take on an unquantifiable significance that would destabilize not only capitalism but Marxism and any other utility-valued system of exchange as well.

What we have here is a clash of two "economies," one measuring value through the personal enrichment of bonding and enworlding, the other measuring value as utilitarian commodity for autonomous self-maintenance. One sees expansive, non-zero-sum pleasure and social bonding in giving; the other sees virtue in the pursuit of zero-sum profit and aggressive private advancement as a positive human trait. Pierre Bourdieu takes this point to its radical end as he discusses Derrida's impossible gift: "[I]t is not possible to reach an adequate understanding of the gift without leaving behind both the philosophy of mind that makes conscious intention the principle of every action and the economism that knows no other economy than that of rational calculation and interest reduced to economic interest."[8] Reason will always be able to convert human behavior to balance-sheet accountability by interpreting any action as a self-interested move to gain some utilitarian edge. Recall, for example, how Oates' altruism could be reconceived as an action bent on winning him honor and glory. And so commodity-exchange economics will always be able to measure zero-sum debt and obligation. Ultimately, it will require the "madness" of escape from rationalist analysis (product of a vision of isolated mind) to accept the value of bonding implicit in the gift ethic. And what must be weighed is whether or not much madness in this case is in fact divinest sense, whether or not our current rational system is sufficient for our continued health. The idea that gifts can be transformative and regenerative returns us to the idea of comedy.

But first we should keep in mind that the binarism of gift versus market economy is itself somewhat problematic. Osteen makes a crucial observation: like comedy itself, "gift practices do not follow rules: they seep outside of our categories" (25), and "the fact that the gift beckons us to think about value both in terms of money and beyond money demonstrates not its incoherence but its elasticity" (34). So, rather than seeing binarism at work, we should perhaps follow the advice of Alain Testart who proposes a scale running, according to Osteen, "from the least to the most coercive forms of sanction" (6). At one extreme are true gifts, "which are acts 'of someone who provides something without demanding a return,'" of which anonymous charitable donations are the highest form (and closest to Derrida's idea of gift). At the other extreme are "transactions that are clearly obligatory [including] . . . creditor/debtor relations in Western societies, where a creditor can seize goods or garnishee wages" (6).[9]

What's a Nickel Worth?

As with the idea of the gift, the idea of comedy slides along a scale, from unfettered celebration of giving found in pure carnival to the rather somber works of contrition, exchange, and debt repayment found in more moralistic, doctrinal, and satiric comedy.

Take the case of David Mamet's 1975 play *American Buffalo*. A junk dealer has sold an American buffalo nickel to a collector for what the dealer thinks is a sizable profit, only to find out that the nickel *may* have been worth more than he received. Here we have a curious case of an object at face value worth only five cents fluctuating in value based on some value-added sense of rarity and relative historical value. This matter brings up the issue of antiques valuation in general. As antique or collectible, an object assumes qualities of its past "life" and frequently of its past owners in a manner similar to what Mauss identifies: a transferal of an actual part of the previous owner to the recipient through the gift. In the collector's world, however, such objects are in turn returned to commodities by being assessed a market value and exchanged (or insured) in an actuarial manner equivalent to raw market commodities. Gift value is transformed to market worth. In *American Buffalo*, the buffalo nickel's value continually shifts depending on the market in which the piece is assessed. From its five-cent face value to a "good sale" by Don the shop owner, to Don "being taken" by the collector, the nickel's brief history on stage reveals a simple, face-value commodity assessment being transformed into a far more complex issue involving exactly how to assess "worth" to that which presumably has value at multiple system levels.

It is worth noting here that that master economist Adam Smith, author of *The Wealth of Nations* (1776), also authored an earlier work entitled *The Theory of Moral Sentiment* (1759), where he distinguishes between justice and virtue: "The rules of justice may be compared to the rules of grammar; the rules of other virtues, to the rules which critics lay down for the attainment of what is sublime and elegant in composition. The one, are precise, accurate, and indispensable. The other are loose, vague, and indeterminate, and present us with a general idea of the perfection we ought to aim at, than afford us any certain and infallible directions for acquiring it."[10] Regarding the point, Eun Kyung Min notes, "[W]e have a clear insistence on the disanalogy between moral and economic commerce. The rules of justice, like the rules of economy, work through an intentional 'grammar' of motives and means. . . . Virtue, on the other hand, follows rules only in the sense of instancing standards and qualities we hold up as examples."[11] Min moves Smith's observations of justice and virtue to the economic plane that Smith would later address by observing that *The Theory of Moral Sentiments* is committed to virtue over justice while the opposite is true of the far more influential *Wealth of Nations*. Put another way, the justice of tightly logical commodities exchange as laid out in *Wealth of Nations* has turned out to prevail over the moral pursuit of virtues found in the far less exacting but far more consequential gift ethos.

Part of what *American Buffalo* demonstrates is the difficulty inherent in a commodities economy of maintaining a sense of justice without encrusting its culture in legalisms that ultimately lose sight of their original *virtuous* designs. Virtue loses its creative flexibility as it becomes judicially and economically systematized. This fundamental disconnect in part explains why such commodity economies are so lawyer dependent, requiring as they do processes of adjudication to at least try to imitate flexible and aneconomic virtue. In the case of this small junkyard community, Teach becomes the "lawyer," filling Don with pseudolegal propositions, justifications, and resolutions. Don feels he's been taken by the collector who buys his nickel, and he concludes that this injustice needs to be rectified even if he needs to appeal to the old standard argument that the ends justify the means. Don's world has fallen out of equitably just and virtuous balance, and it requires adjustment even by countertheft if necessary. Mamet's play masterfully reminds us that wealth accumulation in the Western world is often a process involving ethically suspect and even criminal activity cloaked in socially legal garments.

What we see in *American Buffalo* is how the nickel receives multiple assessments and finally goes beyond economic exchange value into another realm altogether. Teach is a hard-nosed street philosopher who sees everything as devoid of living attributes, simply possessing bald monetary value. Don's down-and-out friend Bobby intuitively sees the world otherwise, though he's incapable of articulating his perspective (even in "inarticulate" Mametian terms). Teach sees Don as having been cheated by a crooked collector and is thereby entitled to "cheat" back. Bobby simply sees Don's hurt and disappointment, which in turn affects their friendship. Bobby in turn struggles to redeem their friendship by recovering the nickel, or a reasonable facsimile thereof. Teach, of course, declaims, "Loyalty does not means shit a situation like this."[12] Business is bottom-line profit and loss. Virtues like friendship and loyalty are of no economic worth.

Bobby struggles throughout the play to be permitted to help Don and Teach retrieve the nickel, although his ultimate goal is to maintain or reestablish a friendship under strain thanks to the lost nickel and Teach's intrusion. In a twist common to market economy, Bobby believes that getting the nickel back will help him to "buy" Don's friendship back. Bobby's heart is in the right place since, though confused, his idea of gift seems to dominate amid all the verbal detritus that Teach spews out. The nickel has become for Bobby a gift he must secure and then give to Don as a gifted sign of friendship. The nickel becomes "gift," as Mauss describes it, of undetermined commodity value carrying with it part of the giver as it is passed onto the recipient, designed to secure and assure a bond of friendship between the two men.

So amid the clutter of dubious commodities in Don's Resale Shop is an extended market-exchange discussion, complete with schemes for recuperating that which has been "stolen" and which has therefore caused disequilibrium in this shop. Though this discussion predominates the play; below and beneath these exchanges occur efforts to use the coin to reestablish a strained friendship. In the end it works: Bobby and Donny are reunited, and Teach is banished. Ironically, no sign of the coin ever appears: we see here clear evidence that it's literally the value of the thought that counts more than any idea of actual retail value. Gift is that part of the giver given, often though not always manifest in a physical property like a coin. The coin, finally, becomes a gift by virtue of what the coin embodies exclusive of or beyond its market value.

Osteen makes the following point: "In a gift economy objects are personified; in a market economy, persons are objectified."[13] Teach and his economics lessons objectify people, determining, for example, Bobby's "value" as a participant in the untried plan to retrieve the coin. Don, nearly convinced of Teach's worldview, ultimately rejects Teach's teachings and accepts the virtued humanity of Bobby's gift offering. That alchemical transformation of the valued buffalo nickel to a valued friendship gives us pause to consider what value exactly it is that can be placed on friendship.

Merchant Gifting

The issue of the gift and reciprocity extends beyond economics, of course, and into matters of justice, as Adam Smith's comments above indicate. Justice, after all, is a sort of austere economics of right/wrong action and consequent recompense. It balances the one against the other and passes judgment on how equity can best be reclaimed following a wrongful zero-sum act leading to some element of social disequilibrium. When the Duke in *Measure for Measure* initially metes out justice, he declares,

> "An Angelo for Claudio, death for death!"
> Haste still pays haste, and leisure answers leisure,
> Like doth quit like, and Measure still for Measure. (V.i.405–7)

The Duke ultimately relents, sparing Angelo and even telling the lecherous Lucio, "Thy slanders I forgive, and therewithal / Remit thy other forfeits" (V.i.514–15), including whipping and hanging. In part, accepting complicity in his city's corruption, the Duke offers judicious, nonzero-sum gifts of remission to his various legal forfeits, disturbing the judicial balance sheet in the process but doing the *virtuously* right thing in this dark but finally comic world and in the end establishing a new order benefiting all the denizens of the Duke's Vienna.

The unbalancing madness of midsummer affects Theseus in *A Midsummer Night's Dream* in much the same way. Preparing to witness the play put on by Bottom and his crew, "A Tedious Brief Scene of Young Pyramus and His Love Thisby," Theseus' war-won bride, Hippolyta, declares that she's not willing "to see wretchedness o'ercharged, / And duty in his services perishing" (V.i.85–86). Gracious, giftful Theseus, however, responds,

> The kinder we, to give them thanks for nothing.
> Our sport shall be to take what they mistake;
> And what poor duty cannot do, noble respect
> Takes it in might, not merit. (V.i.89–92)

Here Theseus exhibits something a bit different from the Duke's actions. Weighing the efforts of Bottom and his crew and determining that the "product" actually is "defective and of little worth," Theseus nonetheless looks beyond the qualities of artistic value. He doesn't evaluate the value of the interlude for its qualitative "market" value, but instead judges the sincerity of the gifters themselves, much as a parent values a child's handmade gift over a more costly purchased present. Theseus sees "value" deriving from his subjects' good intent; parts of them inhere in the gift, and Theseus' kind reception of their gift of play to him is itself a gift generated by the leader's insight and benevolence, the kind of social binder that Mauss speaks of.

So whereas Vienna's Duke offers a gift of forgiveness as a sort of compensation for his previous governing shortcomings, Theseus of Athens offers kindness through a subtle understanding of "value," wherein the value of the subjects' good intent—their gifts of themselves—is manifoldly repaid by the master's good will, which generates value-added social bonding in the process. Prospero, in *The Tempest*, demonstrates yet another level of gift. At one level, his final treatment of Caliban is equivalent to the Duke's, confessing that "this thing of darkness I / Acknowledge mine" (V.i.275–76). Like the Duke and his fallen Vienna, Prospero accepts that he is complicit in the imperfect behavior of Caliban, and his pardon of the creature falls at least in part into the realm of recompense for poor leadership. Of the other offenders on the island, Prospero is far less complicitly entangled in their behavior. After harmlessly tormenting them throughout the play, allowing their guilty consciences to blossom, Prospero tells his near-fratricidal brother, "I do forgive thee, / Unnatural though thou art" (V.i.78–79). Breaking the legalistic circle of reciprocity known as vengeance, Prospero's gift is a non-zero-sum gift expecting nothing in return.

What Prospero aims at is the utopian world Gonzalo dreams of when he first scans the desert isle:

> I' th' commonwealth I would by contraries
> Execute all things; for no kind of traffic
> Would I admit; no name of magistrate;
> Letters should not be known; riches, poverty,
> And use of service, none; contract, succession,
> Bourn, bound of land, tilth, vineyard, none;
> No use of metal, corn, or wine, or oil;
> No occupation; all men idle, all;
> And women, too; but innocent and pure;
> No sovereignty. (II.i.143–52)

In a naturally bountiful world without the need for scales, courthouses and economic institutions have no place. Where bounty is held in common, the need to balance exchange and reciprocity do not exist, and the spirit of unchecked, increasing, and perpetuating gifting eliminates the need for judicial oversight. Gifts in such an aneconomic world become nothing more than, and nothing less than, the bond between humans that make for utopian society itself. Says Hyde, "What is given away feeds again and again, while what is kept feeds only once and leaves us hungry" (21).

Freely giving and receiving eliminates any oppressively individualistic concerns like debt, obligation, or dominion. In fact, the world and how it sustains itself, what it feels to be crucial for sustenance, has changed, and Derrida's impossible gift materializes not as a result of a thus-far unconsidered change in *context*. Rather, if we change from a zero-sum to non-zero-sum existence, ideal gifts would become possible and even commonplace. In other words, if you can find a way to offer an ideal gift, you have found a way to change the world. If humanity can forget notions like debt by stepping beyond individualized consciousness, then perfect gifts will bond perfectly. The two memic discoveries go hand in hand.

This, of course, is the stuff of lunatics, lovers, and madmen, the kind of people captivated by a hopelessly optimistic spirit of comedy, unattainable and therefore little worth considering except as an idle pastime for the unproductive and parasitic. This is actually the kind of charge made against Marcel Mauss, whose romantically nostalgic vision of how to change early twentieth-century Western culture has been met with cynicism, much like Bakhtin's carnivalesque vision of change has been charged with being naïvely romantic. But the cynic's response surfaces at least in part because the comic's vision is so powerful. As Osteen suggests, "[T]o discover the true nature of the gift, we must redirect our gaze from reciprocity toward other principles and motives. When we do, a different set of norms emerges, a set founded upon spontaneity rather than calculation, upon risk instead of reciprocity, upon altruism instead of autonomy" ("Introduction," 7). Osteen identifies terms attachable to Testart's continuum, spontaneity/risk/altruism at one end and calculation/reciprocity/autonomy at the other. Gonzalo's vision, not unlike Mauss' romanticism, is hopelessly romantic if for no other reason than that the calculating scales of economy are currently so centrally situated in the Western human condition. As a result, it is virtually impossible to eliminate the one extreme of scarcity-induced, nonzero commodity exchange and give life to a world of non-zero-sum bountiful gift exchange. Impossible, that is, except in the world of comedy. But even then the specter of "commodity exchange" often surfaces.

Recall, however, that Bakhtin's analysis of Rabelaisian humor sees the carnivalesque rising from the market itself. At such gatherings, festive indulgence celebrating earth's bounty allows for occasions of luxury, overabundance, and even waste. They are events that mark the world of gift in general. But we must not forget that the market is site of culturally oversighted commodity exchange, a fact that during festive gatherings provides for an interesting mix of both exchange extremes. In times of scarcity, doubtless, markets align more fully with the dominant culture's more controlling and "efficient" commodity function; in times of abundance, the unreciprocating potential of gifting is allowed greater liberty along with greater robustness, which in turn undermines the zero-sum circle of the agelast's closed society of which commodity markets are central. Carnival is community's gift to dominant culture, dependent on bounty and always on the verge of collapse back into the order of precise zero-sum exchange and its hallmark oppressive preservation and distribution of wealth. However, when gift/comedy/carnival take center stage, dominant culture's commodity economy—and the closed social order it feeds and manifests—is forced into the shadows.

The Merchant of Venice is Shakespeare's consummate study of gift versus commodity exchange. Portia's touchstone "quality of mercy" speech is central to the play and, more significantly, to comedy in general:

> The quality of mercy is not strained;
> It droppeth as the gentle rain from heaven
> Upon the place beneath. It is twice blest;
> It blesseth him that gives and him that takes.
> 'Tis the mightiest in the mightiest; it becomes
> The thronèd monarch better than his crown.
> His scepter shows the force of temporal power,
> The attribute to awe and majesty,
> Wherein doth sit the dread and fear of kings;
> But mercy is above this scept'red sway;
> It is enthroned in the hearts of kings;
> It is an attribute to God himself,
> And earthly power doth then show likest God's,
> When mercy seasons justice. (IV.1.182–95)

Mercy blesses the receiver, of course, but it returns to the giver as well in a non-zero-sum fashion that oppresses no one as it enriches everyone—a true comic sentiment.

What occurs in *The Merchant of Venice* clarifies the process of gift for Shakespeare's canon in particular and for comedy in general. Using Hyde's distinctions between gift and commodity exchange, Ronald A. Sharp sees the play as a perpetuating flow of gift exchange leading not to zero-sum equivalence in reciprocity but to non-zero-sum increase and therefore to virtually endless bounty. Drawing from Hyde, Sharp observes, "The dialectic of giving and receiving is central to gift exchange. A commodity, however, 'is truly "used up" when it is sold because nothing about the exchange assures its return.' The curious thing about gift exchange is that it always assumes a return but makes it imperative that that fact not have the status of an expectation. The gift 'joins people together. It doesn't just carry feeling, it carries attachment or love,' and its movement is circular."[14] Here Sharp is using (through Hyde) Mauss' conception of gift as social exchange, placing his understanding of gift on Testart's running scale somewhere between pure gift (with no expectations of any return) and commodity (with explicit expectation of full reciprocity). At this point there's an opportunity to update Sharp (and Mauss and Hyde) a bit. Since with the gift goes part of the giver, Alain Caillé's image of a spiral may be more appropriate than the oft-discussed circle.[15] Osteen summarizes Caillé's point, noting that "Caillé suggests that, since the original gift introduces something that was not there before, it cannot be *merely* reciprocal. Further, reciprocity explains why things are given back, but not why one would give back more than one receives" ("Introduction," 7). The "more" here at least in part includes the element of the giver's "self" that inheres in the gift as it is given and then passed on. And as more is given, the circle breaks into a rising spiral of gain for all involved.

Sharp provides an extensive list that demonstrates that gifts and giving abound and increase in *The Merchant of Venice*, from the gifts Antonio gives to Bassanio and gifts Bassanio gives to Portia to the numerous exchanges of rings throughout the play. Included among Sharp's long list of tangible exchanges should be added the instances of Portia, Bassanio, and others exchanging their love with each other, cases of intangible and even risky gifting.

Most curious among the list are those including Shylock, the fly in the play's comic ointment. But recall that Shylock at least claims to *give* Antonio the money that Bassanio needs in exchange for a virtually worthless promissory note of a valueless "commodity," Antonio's pound of flesh:

> I say
> To buy his favor I extend this friendship.
> If he will take it, so; if not, adieu.
> And for my love I pray you wrong me not (I.iii.163–65).

That this might be subterfuge is, of course, a concern validated by Shylock's later sinister actions. That at play's end Shylock gives "gifts" to Jessica and Lorenzo only because he is forced to is pretty clear. And the Duke's "gift" of pardon for Shylock at play's end is hardly one well received. But a touchingly human moment wherein Shylock does lie within the realm of gift's influence involves the lost ring Shylock longs for, given to him by his now-dead wife. When he hears that Jessica has exchanged it for a pet monkey, Shylock cries out, "Out upon her! Thou torturest me. . . . It was my turquoise; I had it of Leah when I was a bachelor. I would not have given it for a wilderness of monkeys" (III.i.107–9). Clearly, Shylock is aware of the more-than-commodity value of objects, though this gift was given decades before the opening scenes of this play and his survivalist instincts have turned him into the most autonomous of creatures in this mercantile world driven by contracts and formally legalistic procedure.

Shylock is, of course, the agelast in the play. He is the force against which comedy must struggle, the radical personification of individualized human appetitiveness that current society sees as crucial to its prosperity. In one manner of looking at matters, Shylock actually represents Venetian order itself, albeit in its negative, raw-in-tooth-and-claw manifestation. He has effectively converted the highly impersonal mercantile ethic of reciprocity literally into a flesh-and-blood exchange, mastering the process of contractual obligation and elevating it to a retaliatory game with mortal consequences by ensnaring his nemesis Antonio into putting up a pound of flesh as legally bound collateral. In a complex twist, Shakespeare recognizes that the appetitiveness that dominates Shylock is a force that must be given some reign in order for society to prosper. Even the forfeit Antonio sees the necessity of accepting the legal system that ensures Venice's mercantile prosperity:

> The Duke cannot deny the course of the law:
> For the commodity that strangers have
> With us in Venice, if it be denied,
> Will much impeach the justice of the state,
> Since that the trade and profit of the city
> Consisteth of all nations. (III.iii.26–31)

But there is far more to living than mere profit. And there's a point at which even contracts of honor must be challenged, a point where orderly justice is in need of being trumped by the disorderly invasion of virtue. But then again, in a world founded on impartially legalistic justice—economic and otherwise—there must be a way that justice can remain intact *and* that virtue can find its place.

Portia enters Venice as judge, loaded with the "wisdom" of love and filled with the noneconomic non-zero-sum virtue of mercy—a true loss leader, to use economic jargon—and her entrance suggests that Venice is on the verge of significant transformation. Ingeniously, though in line with comic expectation, Venice's laws of justice actually become a means of manifesting the new Venetian order that Portia envisions, one that will acknowledge the need for material comfort but only inasmuch as it secures a foundation for the *higher end* of remaining virtuous. In this case, the just obstructionism of the law converts to the salvation of justice as Portia manipulates the law to save Antonio and condemn Shylock. From that point Portia additionally encourages invocation of the quality of mercy to spare Shylock of his "just" doom, leading to a triumph of virtue overall. As Sharp notes, "So long as they [forms and conventions] are life-giving, so long as they avoid the sterile legalism and narrow vision of Shylock, forms serve the highest values, including human intimacy" (264).

What we are seeing is that comedy at its highest point is fundamentally pre- or post- (certainly non-) capitalist and pre- or post- (and certainly non-) Enlightenment. This brand of comedy challenges the idea that happiness and freedom are ideals to be pursued by individuating methodologies. It favors instead the pursuit of more perfect forms of happiness and freedom achieved (perhaps paradoxically) by communal recognition of virtue and mercy, which need to find their way back into justice even to the point that they disrupt orderly reciprocity.

Shylock, of course, does not freely accept his gift of enforced conversion and subsequent pardon, seeing the gift and community he's forced to join as a severe restriction on his previous legally conferred autonomy. Gifted for his own good, Shylock begs to differ. With Shylock, we see the limits of the gift's potential: it must be freely accepted because it invariably enjoins us to put down our defenses and engage in a community dedicated to the betterment of others. As we dedicate ourselves as gifted participants, we enter the perpetual risk that this community—lacking any formal checks and balances whatsoever—will *not* provide a bounty of good greater than could be gathered via individual pursuit of commodity exchange. The risk of undoubted individual loss in return for suspect communal gain is clear, and it's one that Shylock prefers not to take, competent as he is to maintain autonomy as a money lender under protection of Venetian laws and preferring to remain connected to the loose confederacy of other money lenders in Venice's Ghetto.

Submitted to Dawkins' sucker-grudger-cheater model, Shylock fits pretty comfortably into the category of cheater, more than willing to see his vulnerable, sucker customers play into his self-interested (and socially sanctioned) schemes of personal gain. Portia disguised as legal advocate reveals herself to be something of the grudger that Dawkins describes in his model. She finds a way to top Shylock and undermine the entire tight legalism that allows ingenious cheaters in general to manipulate the system and flourish in Venice, especially amid altruistic suckers like Antonio. But if this is an example of a well-armed grudger's victory over a cheater, it's an example with only limited hope for long-term change. After all, what Portia accomplishes amounts to little more than the recovery of one victim, Antonio, from a system that remains pretty much intact.

But there is another part of *The Merchant of Venice* that isn't set in Venice proper—its locale being Portia's inherited estate Belmont. In Belmont, Portia is faced with a slightly different problem. Early in the play, Portia gives her love to Bassanio, giving him a ring as token of her undying commitment. In Venice, however, Bassanio is forced to relinquish the ring as payment for "services rendered" by the mysterious Doctor of Law (the disguised Portia). Once back in Belmont, Bassanio is vulnerable to charges of defection, but after mildly chastising him, Portia once again extends the cooperative quality of mercy and forgives to the defenseless and repentant Bassanio. This sort of happy ending is not particularly original. What is curious, however, is why Shakespeare chose to yoke together these two seemingly unrelated plots. Perhaps, most curiously is why Shakespeare used two distinct locales at all. Couldn't the love story have occurred within a palazzo in Venice proper?

Osteen observes a significant tenant of cooperatively informed gift economy relevant to this question, namely that in a world otherwise dominated by commodity exchange, "the gift is an economy of small groups" ("Introduction," 17). There is a clear vulnerability attached to exposing oneself by extending the gift beyond a relatively small, trusted community. Shylock's own self-interested resistance reminds us that Venice in general may slip into moments of sentimental relief that Antonio is spared, but the city will shortly return to its former ways. Where a community based on the gift can survive and even possibly thrive, it appears, is in a protected green zone like Belmont, far from mercantile meccas like Venice. This observation by itself virtually condemns comedy to the hidden recesses beneath or beyond dominant culture and reminds us that retreating to Belmont is the sort of escapism that comedy frequently endorses. But it's also the kind of withdrawal comedy must resist or else risk becoming (or remaining) irrelevant.

Taking a cue from chaos theory, game theory asks that we consider what chaotics calls sensitive dependence on initial conditions: everything depends on where we begin. In real-world terms, cheaters currently dominate as a result of individuated cultural and natural inclination. The hope, according to game theory, as summarized by Dawkins, is that if the more rare, guardedly reciprocating altruists build strength "by cooperating with one another in cosy little local enclaves, [they] may prosper so well that they grow from small local clusters into larger local clusters" (219). As they strengthen under mutual protection, "they may grow so large that they spread out into other areas, areas that had hitherto been dominated, numerically, by individuals" (219) who look for every opportunity to cheat.

Two points make this scenario something more than a utopian fantasy. First, it is not a case necessarily that there must be a conscious conspiracy to group together given that guarded altruism naturally thrives among itself, meaning that kindred spirits will naturally congregate in the very way they did at Portia's Belmont. Recall, for example, how effectively the two self-interested suitors and their entourages were cast off the island, thanks to the suitors' contest designed by Portia's deceased father, the play's first guarded altruist. And secondly, as is also seen in Belmont, guarded altruism has a robustness that allows altruism to thrive, given that all the members are frankly willing to see others succeed without feeling a need to compete for the status of first among many. Belmont, from this point of view, is not some sanctuary built to hide from the world, but rather is a breeding ground for guarded altruists to build a critical mass and then perhaps return to Venice to finish the job Portia had begun during her guerrilla infiltration into that mercantile epicenter, proving the comic vision of guarded altruism to be viable but requiring larger numbers, finally, in order to prevail.

The Critical Mass of Gifting

In Moss Hart and George S. Kaufman's 1936 comedy, *You Can't Take It With You*, Belmont moves from the outskirts of sixteenth-century Venice into the twentieth-century hustle and bustle of New York City, and Portia transforms into an Aristophanic old man transformed into rejuvenated youthfulness. The play is at once whimsical and realistic in its approach to the challenges of the world. Grandpa, in conversation with his high-strung Russian exile friend Boris Kolenkhov, sums it up quite nicely:

> GRANDPA: Oh, the world's not so crazy, Kolenkhov. It's the people *in* it. Life's pretty simple if you just relax.
> KOL: . . . How can you relax in times like these?
> GRANDPA: Well, if they'd relax there wouldn't *be* times like these. That's just my point. Life is kind of beautiful if you let it come to you. . . . But the trouble is, people forget that. I know I did. I was right in the thick of it . . . fighting, and scratching and clawing. Regular jungle. One day it just kind of struck me, I wasn't having any fun. . . .
> KOL: So you did what?
> GRANDPA: . . . Just relaxed. Thirty-five years ago, that was. And I've been a happy man ever since.[16]

"Let life come to you" is the central message of virtually *all* comedy, romantic and otherwise. It's a message that from an agelast's perspective is indeed pretty crazy, a point that is the root of this particular play's humor, populated by eccentric but lovable personalities lost in their obsessions without any real care for anything but personal contentment amid interpersonal harmony. The "relaxation" they variously pursue helps them utterly escape the rat race of the world of commodity exchange, leaving every contact they make fall into the domain of gift exchange. And the small community they create even within the commodity's capital of the Western world is one that demonstrates the small successes of comedy when it achieves even a modest critical mass within a world of capitalist cheat and oppressive self-interest.

From an orthodox perspective, one would look on this band of eccentrics as little more than a band of hopeless but harmless losers unable to survive in the "real" world. But it's Grandpa, the originator of his family's eccentricities, who began this journey to un-self-conscious contentment. Significantly, it's a choice Grandpa made thirty-five years ago not because he was a failure in the world of commerce but because he had mastered that game and tasted its emptiness: "I used to get down to that office nine o'clock sharp no matter how I felt. Lay awake nights for fear I wouldn't get that contract. Used to worry about the world, too. Got all worked up about whether Cleveland or Blaine was going to be elected President—seemed awful important at the time, but who cares now?" (75). Apparently having invested enough capital to create a haven of at least moderate bounty, he accumulated enough to resign from that world with sufficient guarded reserve to eliminate any sucker's vulnerability against a potential onslaught from the outside world.

The representative capitalist of the play, Mr. Kirby, initially calls Grandpa's philosophy "dangerous," "un-American" (75), and "downright Communism" (76). However, though trained to follow in his capitalist father's footsteps, Mr. Kirby's son, Tony, is caught in the gravitational attraction of this comically free world and wants to marry into this family, turning his back on the family business into which he is born in order to become part of a "real family" (77) given to love and simple contentment. Toward play's end, Tony reminds his dad of all the youthful dreams his father once had (revealed in letters serendipitously discovered from dad's youth), including becoming a trapeze artist and saxophone player. Mr. Kirby responds, "I may have had silly notions in my youth, but thank God my father knocked them out of me" (76). Listening to Tony argue with the agelastic Mr. Kirby, Grandpa observes to resistant Mr. Kirby, "I think if you listen hard enough you can hear yourself saying the same things to *your* father twenty-five years ago" (77). Grandpa continues, "We all did it. And we were right. How many of us would be willing to settle when we're young for what we actually get? All those plans we make. . . . what happens to them? It's only a handful of the lucky ones that can look back and say that they even came close" (77).

Ultimately, even Mr. Kirby gives in to Grandpa's comic-spirited siren call, taking up the saxophone and granting his son permission to marry into this most unusual family. Mr. Kirby joins the fireworks artisans, playwright/painter, ballerina, printer, actress, servants, revolutionary, and blintz-making exiled Grand Duchess in a home that the outside world of care has forgotten.

Actually, the outside world hasn't forgotten them. As can be expected from the game-theory models above, it would at first glance seem apparent that agelastic interest in destroying this enclave of comic resistance would be overwhelming. In fact, the threat to this household is omnipresent, far more threatening than, say, a Shylock would be, since the threat here arises in the form of the U.S. Tax Service. Grandpa has *never paid income taxes* and is in jeopardy of losing everything to this outside agelastic force of conformity. Undeterred, or more particularly unconcerned, the family carries on as (un)usual even as it receives warning letters from the government that are variously lost and found. It's a menace of certain doom that lurks behind the play almost from the start and appears to have no solution. The world of exchange nearly engulfs the stage. But the generosity of these people, giving homeless and deserted denizens of the neighborhood shelter, food, and uncritical affection, remains intact, without expectation of return whatsoever. The bounty of joy and love received actually immeasurably outweighs what is given. Even an apparently undeserving Mr. Kirby is uncritically embraced and absorbed into this tantalizing existence of loving self-fulfillment. Perhaps *he* will repay the gift of kindness by bailing the family out of trouble.

This option, however, is never presented. Rather, the family is saved by what could actually be considered impossibly close to a Derridian act of "pure" gift. The audience is told that several years ago, a deliveryman came to the house, fell in love with its aimless generosity, blended into the fabric, and never left, actually eventually dying on the premises. The family gave him a proper burial, never knowing his true name. In an act symbolic of utter generosity lacking any bit of hope of reciprocity, Grandpa gave the man his own name to be buried under. The stranger was known and loved as a member of the family, and the name-giving gift merely warrants a footnote in the family's catalogue of unorthodox but kindly intended behavior. In an unhurried manner belying any sense of concern, the play reveals an odd penultimate event. Grandpa receives a letter of apology from the Tax Service: "I don't owe 'em a nickel; it seems I died eight years ago" (79).

Grandpa's gift of himself to the stranger returns to him in a spiral that provides a bounty of unexpected joy. They are now and forever tax free, unbound to the outside system of control, and free/open to influence and convert others as they have converted even the captain of industry, Mr. Kirby. The "return" unexpectedly received as a result of Grandpa's gift to the dead is the even greater gift of an escape from agelast entrapment, guaranteeing continued uninterrupted joy in the world for this family and all who are spun into its circle/spiral of giving. Thanks can't be given to the dead stranger, and the stranger is not even aware of the gift he's given "in return" (at least not in any conventional manner)—hence, Derrida's "perfect" gift.

Before the play actually closes with a resumption of benignly comic orderly disorder, the clan settles down to dinner and to thanks voiced by Grandpa:

> Well, Sir, here we are again. We want to say thanks once more for everything You've done for us. Things seem to be going along fine. Alice is going to marry Tony, and it looks as if they're going to be very happy. Of course the fireworks blew up, but that was Mr. De Pinna's fault, (DE PINNA *raises his head*.) not Yours. We've all got our health and as far as everything else is concerned we'll leave it to You. Thank You. (80)

Grandpa gives thanks to the prime Giver for that initial, singular, non-zero-sum gift of life/living. Having found how to accept life as it's offered is typically the comic domain of youth or of simple folks. Grandpa is neither young nor simple, but he has adapted his biology to be as much of both as possible. This Aristophanic resolution wherein age rediscovers the fountain of youth is one of the play's central accomplishments. More shrewd than anyone in the household, Grandpa is clearly among the happiest because of what he has in contrast to what he once was. Mr. Kirby is on the way; he is fully embraced by the clan, guaranteeing that his body's perpetual indigestion will shortly be a thing of the past despite the rich/bountiful diet of goose and blintzes he's about to receive and consume. His body tells him what his mind has ignored since his youth; fortunately for him, this household has awakened his mind to listen to his body.

In *You Can't Take It With You*, harmony is achieved not by naïvely hoping that the world/environment will magically/mythically change itself. Rather, harmony is achieved by shrewd awareness of a world of cheats and adapting to it through guarded oversight and continued openness to those caught up by its tantalizing allure and the bounty that such generosity in turn returns. The triumph involves acceptance, tolerance, and gift. And it also paradoxically argues that individuals triumph best when they become community. It's a triumph limited to a small group within the larger circle of society, but it is growing. The group accepts its fragile quality, living as it does within the belly of a larger world of exchange, but it doesn't expose itself needlessly or unprotectedly to the crushing potential of outside forces. Finally, *if* change will occur, it appears that it will happen one individual into one small community at a time.

All for Love: The Gift of Marriage

Comedy at its hardnosed best reveals several crucial points. That the world is populated mostly by agelastic "cheaters" intent on satisfying personal needs is a given, evidenced most strongly in works like the two operas of Gay and Brecht. But perhaps more significantly, it reveals that unilateral altruism—the agenda of much naïve comedy—is a subcategory of empty and fruitless idealism, unsupported by materialist reality and ultimately even subject to tragic rather than comic resolve. The idea of behaving in a unilaterally altruistic fashion is one seemingly destined to failure. Comic protagonists must seek cooperation while conceding that attendant vulnerability must originate from a position of guarded strength. Being willing and able—though disinclined—to fight for peace is, rather paradoxically, the hard-nosed comic message.

A related paradox reveals itself in comedy involving the paradoxical institution of marriage. Protestations of love and tokens of affection flood the comic stage throughout its long history. But when marriage is allowed to dominate the stage as an absolute good, comedy has failed in its larger function as community builder. *The Merchant of Venice* is an excellent case in point of the challenges attending exposure to the vulnerabilities of love turned into marriage. The vulnerability that love-as-marriage compact generates is exposed when Bassanio is forced to relinquish Portia's ring, against his promise to his betrothed, in exchange for services rendered by the Doctor of Law (the disguised Portia). His keen reluctance to betray his vow, however, is overcome by his promise to give anything the doctor asks for in thanks for having saved his dearest friend, Antonio. Rather ironically, his reluctance to give the ring to the doctor threatens to undermine one of the chief tenets of gift exchange, namely that it be circulated among an ever-growing circle/spiral of worthies. Caught between two commitments, Bassanio explains his reluctance:

> Good sir, this ring was given me by my wife,
> And when she put it on she made me vow
> That I should neither sell nor give nor lose it. (IV.i.439–41)

To this defense Portia disguised as the doctor replies,

> That 'scuse serves many men to save their gifts.
> And if your wife be not a mad woman,
> And know how well I have deserved this ring,
> She would not hold out enemy for ever
> For giving it to me. (IV.i.442–46)

And the disguised Portia's bidding is further supported by Antonio:

> My Lord Bassanio, let him have the ring.
> Let his deservings, and my love withal,
> Be valued 'gainst your wife's commandèment. (IV.i.447–449)

The market value (Bassanio notes that "[t]here's more depends on this than on the value" [IV.i.432]) is not the rub in this transaction; rather, it's the matter of giving away the value added to the ring by the fact that the ring is "filled" with Portia's love. Ironically, however, that value would be much diminished should Bassanio refuse to let the circle grow by refusing to pass the ring on to the worthy doctor, especially given Antonio's behest.

By giving it to the doctor, the ring's gift value increases, its legacy as a value-added gift moving with it into the possession of the doctor. Bassanio's solemn vow to never give up the ring, broken under duress by the doctor's just demand for it, is something Portia can legalistically convict Bassanio of violating. However, Bassanio's legalistic violation paradoxically conforms to the virtuous code of gifting in that giving not only preserves value but actually enhances it.

If we are to condemn Bassanio for violating his vow to Portia, it would be because it is essentially a violation of a bond rather rigidly built into the institution of marriage, a point that suggests something of the problem implicit in that institution. Being as it is in large measure a legal commodification of love, the institution legalistically restricts and controls the virtuous gift circle of love to a commodity undertaking. Love as gift freely given becomes love with strings of restrictive exclusivity attached.

In other words, the marriage contract secures love by judicially over-sighted sanction, jeopardizing its free distribution as gift. It becomes a major institutional bump along the path of otherwise unrestricted gift exchange, particularly when the exchange involves a man and a woman. Wrapped in the bonds of legalistic matrimony, gift exchange bumps into institutionalized limitations: the free exchange is restricted to a community of two. In *The Merchant of Venice*, however, this problematic limitation is circumvented, given that Portia first forgives Bassanio's transgressions and then willingly includes Antonio into the circle, extending a gift to Antonio *and* to Bassanio unnecessary by law but fully endorsed by the value-added spiral of gift exchange. She "takes" her husband by legal right while giving him his friend as extralegal bounty. Love manifest by forgiveness invades the circle of legalistic marriage, transforming that culturally prescribed bond into an interaction predicated on virtuous inclination rather than judicial, legalistic, or economic obligation. What happens is that marriage transforms from legal document to natural union, but only after a disorderly trial by fire. And the natural union erupts into an opening of the circle to include Antonio, orbited itself by the bond between Nerissa and Gratiano who have undergone a similar love trial. Far more than a selfishly motivated legal bond prevails. And resolution through guarded altruism comes one step closer to fully inclusive fruition.

Note that in *Twelfth Night* this conclusion is also a possibility. Sebastian's best friend, curiously also named Antonio, has very nearly sacrificed all for his companion; but when the rings of marriage conclude the play, Antonio is left out of the cycle. The play rather surreptitiously reminds us that for all its concluding celebration, marriage can sometimes be a selfish commodification of the gift of love that legally endorses restricting the flow of bounty's love. In its purest form, orderly marriage overridden by disorderly love changes the relational bonds between man and woman by taking the gift of love and growing it for ever-increasing bounty. But that potential for increase, strictly speaking, unfortunately often finds a limit in whom it can benefit. As such, the marriages in *Twelfth Night* do not end as "comically" as those in *The Merchant of Venice*. The bonding potential of married *couples* thrives in both plays, but only in *Merchant* is a "third wheel" incorporated, admittedly subordinated to the husband/wife bond but incorporated nevertheless. Portia goes beyond the wildest expectations of a partner in marriage and expands the circle by giving Antonio the ring to give to Bassanio, bringing Antonio into the circle/ring precisely as his selflessness deserves and despite what Portia is legally *required* to do. In fact, Portia offers gifts beyond any expectation, telling Antonio,

> Unseal this letter soon;
> There you shall find three of your argosies
> Are richly come to harbor suddenly.
> You shall not know by what strange accident
> I chanced on this letter. (V.i.275–79)

Curiously knowledgeable of the contents of a "sealed letter" and aware that Antonio's argosies were all reported lost at sea, as everyone knows, Portia must clearly be the "giver" of this bounty to Antonio. This circle of giving continues its spiral of bounty.

A quick look at *Much Ado About Nothing* reveals exactly how spectacular this tale of Portia, Bassanio, and Antonio is. Recall that once Benedick and Beatrice glumly confess their love for each other, Benedick begs to be sent on a quest to prove his love: "Come, bid me do anything for thee" (IV.i.284). Beatrice replies, "Kill Claudio" (IV.i.285), a demand for vengeance on Benedick's best friend whose slander has stained and publicly "killed" her best friend. Though a highly dramatic demand, which Benedick ultimately accepts but fails to fulfill, to a lesser degree it is the stereotypical demand of most soon-to-be-married couples: prove your love by giving up your friends and all you hold dear before I concede giving myself to you. In its most myopic manifestations, marriage reduces love to such obligations of the closed circle, restricting the bounty of love to a gift exchange involving two parties. In the worst of circumstances it actually reduces love to a commodity governed by strict reciprocity. *Much Ado About Nothing* touches on this sanction of a closed circle of two individuals. But then it allows for two separate couples, Benedick/Beatrice and Claudio/Hero, to reunite as a circle of two united couples. However, the circle extends no further, as Benedick tells matchmaker Don Pedro, "Prince, thou art sad. Get thee a wife, get thee a wife" (V.iv.120). A third couple may be invited into the circle, but no third wheel. Two by two is the sanctioned norm. This world bears only a shadow of a resemblance to what Belmont has become.

In the plays above, several appropriately limiting exclusions do occur, most notably in the characters of Shylock, Malvolio (in *Twelfth Night*), and Don John the Bastard (in *Much Ado About Nothing*), characters who ultimately choose and even prefer exclusion. More problematic are the Antonio characters, reminders of the imperfection of the institutionalizing efforts of marriage restrictively controlling and therefore commodifying what is basically uncommodifiable. It's equivalent to marketing carnival, to absorbing and controlling the uncontrollable. Closing the circle and eliminating the potential for growth restricts gifting's elasticity and future capacity for full cultural transformation. This is why, of all Shakespeare plays, we are perhaps most disturbed by Kate's words in *The Taming of the Shrew*, reminders that should the spirit of gift evaporate in a marriage circle, the circle will fast become nothing more than unsavory legalistic obligation:

> Thy husband is thy lord, thy life thy keeper,
> Thy head, thy sovereign; one that cares for thee
> And for thy maintenance; commits his body
> To painful labor both by sea and by land,
> To watch the night in storms, the day in cold,
> Whilst thou li'st warm at home, secure and safe;
> And craves no other tribute at thy hands
> But love, fair looks, and true obedience—
> Too little payment for so great a debt. (V.ii.151–59)

Comedy sees marriage as the beginning of something more than obedience and bigger than a series of debts to be paid. If debts are paid in the strict spirit of reciprocating obligation, then marriage significantly loses its comic appeal altogether, even if it implies mutual security. This reminder by Kate—often happily interpreted as ironic—is perhaps necessary to *The Taming of the Shrew*, though one can hope Kate and Petruchio have found the common ground necessary for true love. But it could also be a cautionary statement tacked onto virtually any comedy ending in marriage. It's not precisely the lesson one wants at the end of a comedy, but it does lurk behind all such endings.

Finally, what we see is that the comic reward for abandoning selfish, zero-sum competition is a non-zero-sum bounty that will feed even our most selfish of appetites beyond their wildest dreams. But it takes a comic resolve that requires insight into human nature greater (and more timely) than Hamlet could gather, and it requires a guarded, inner strength greater than even Troy's Hector could muster. The comic agenda of stooping to conquer (to borrow Goldsmith's phrase) requires all the talents, stretched to their limits, of the best among us.

CHAPTER 7

Comedy Confronts Commodity via the Adaptive Unconscious

That humanity is genetically inclined toward myopic, individualized self-interest is a virtual given. That we can potentially memically alter our genetic inclinations is one great human hope. But against that hope is the point that generating such alterations through some rationally motivated collective act of will is an untenable strategy given that no matter our noble intentions, selfishness in the form of individual opportunism will always hover menacingly above selflessly altruistic attempts to improve the survivability of humanity as a species.

What we do see as a positive possibility against such grim likelihoods, however, is that this longed-for human transformation to speciated, cooperative behavior could possibly, some day, win the field if attached to a strategy that recognizes and creatively confronts the fact that human nature has a selfish, self-serving dimension that needs somehow to be accommodated. Comedy must strive to provide a better alternative even for the cheats and deceivers who choose to take advantage of the world around them for self-interested purposes. Chekhov's 1903 comedy, *The Cherry Orchard*, demonstrates, as many dark comedies do, that selfish commodity exchange has the power to overwhelm the far more fragile value "systems" of unguarded generosity. Even though the businessman Lopahin romantically realizes that market price does little to assess the kind of worth found in the value-added nature of the estate he's just purchased, his singular realization is not enough in this case to save the cherry orchard from corporate development. It is the seeming fate of all products of "sentimental" value when frontally exposed to a system of commodity exchange. Though fighting toe to toe against the seeming inevitability of commodity markets is a poorly advised strategy, looking helplessly on and hoping for a miraculous transformation of the human spirit is just so much untenable human foolishness.

The Gifts of Wit, Humor, and the Adaptive Unconscious

Sometimes, of course, comedy does make itself out to be nothing but so much simple-minded foolishness. But underneath such appearances, challenges do occur, often as a flanking rather than over-the-top maneuver. Among the best such examples is Oscar Wilde's highly entertaining and seemingly inconsequential 1895 masterpiece, *The Importance of Being Earnest*. As the subtitle implies, this "Trivial Comedy for Serious People"[1] combines and confuses the truly trivial with shadowy suggestions of serious import. In other words, often enough comedy's festive dimension belies its more substantial agenda. Striking through the seemingly impervious fortifications of agelastic, dominant culture is by no means an easy task, but it's a task incumbent on comedy to undertake. And even Wilde the aesthete seems to have answered the call.

Alexander Leggatt notes that *The Importance of Being Earnest* is a "listening post in which we can overhear some of the real concerns of late Victorian England."[2] Leggatt sees real issues, such as capitalist concerns that land is no longer a profitable investment (resulting in a shift of power in England), that the lower classes are "breeding" beyond supportability, and that upper-class philanthropy has become fashionable but of no truly lasting value. Michael Gillespie sees similar issues arise under Wilde's cloak of humor. For example, Gillespie notes of the play's manservant,

> Anglophilic P. G. Wodehouse enthusiasts may be quick to see Lane as merely a charming avatar of Jeeves. That sentimental English context leads to the assumption that Lane's tact protects Algernon from having to admit to drinking too much champagne. When I consider the characters as operating out of an Irish identity defined by power relationships, I become attentive to the implications of Lane's dismissive gesture of marriage and of the "young person" who had been his wife. His motives or acts hardly seem benevolent, for Lane's callousness replicates the harshest aspects of the power relationship already established between himself and Algernon, while excluding all its meliorating features.[3]

Buried under puns, inversions, and general apparent non sequiturs are matters of oppression and power politics more overtly presented by many of Wilde's more politically motivated compatriots. However, Wilde's comic design is not to resolve the tensions by rationalist exposition resulting in explanatory closure so much as it is to expose apparently innocuous oppressions hidden behind or under the playful opulence. We may choose to fiddle while the empire smolders, but at the very least comedy will put the smolder on stage to be sniffed out amid the luxury. A pre- or non-rationalist mechanism is at work designed to trigger the beginnings of transformation through laughter and wit, pressing itself against rationalism's orthodox and agelast nose, not quite bloodying it, but working to put it out of joint.

But social satire is not central to this work, certainly not in the way it is central to Wilde's other work. Rather, what more significantly occurs in *The Importance of Being Earnest* is an exercise that goes beneath the subtle social commentary that underlies the play's fun. The schizophrenia of Jack's country and city personalities is resolved in the end of the play by the discovery of his true lineage; and Jack's split personality is allowed to mend, for inside Jack's concocted fictional identity, impossibly, lies truth itself. Fiction and reality, inhering unknown in Jack's former schizoid fiction, fuse: Jack discovers that legally speaking he is Earnest after all. The world, it seems, has rewarded Jack for his wit and ingenuity by bestowing facticity on to what he previously thought to be mere concoction. The either/or of rational consciousness has been replaced by the possibility of "chaotic" and simultaneity permitted beneath the conscious operations of the discretely individualized self. That "reality" catches up to Jack's fiction is less important for Wilde than it is that Jack's fiction is as vital as any mere "truth" that may impose itself on living. Jack lives, and that's what counts. It's actually what is necessary before any hope for transformation of the social fabric can begin.

In "The Rhetoric of Temporality," Paul de Man describes irony as something produced not between discrete subjects but within consciousness, between two selves within a single subject.[4] What we have is a case of self-difference, of an "other" inhabiting a single skin that must be oppressed in the same fashion that external others must. If we take de Man's description and apply it to comedy, we see that part of the general comic charge is to call attention to orderly agelastic obsessions *within* our own consciousnesses and to challenge those influences with something like comically disorderly immunizing antibodies. This point seems to describe Wilde's agenda, seeing as he does that the problem of oppression and dissimulation is frequently less a case of external forces pressing down on personal virtue than of internal forces—however generated—at odds with themselves. Through the capacities to disorient reason with wit, laughter, and irony, comedy often allows the oppressed consciousness or self to surface. And aligning these often-conflicting selves more harmoniously begins to address the matter that oppression is an internal force requiring first-level attention *prior* to moving onto the more obvious social issues that often attract our conscious eye.

Speaking of this sense of divided self, psychologist Timothy D. Wilson draws a further crucial distinction. In his 2002 study *Strangers to Ourselves*, Wilson observes, "The difference between self-revelation and self-fabrication is crucial from the point of view of gaining self-knowledge. . . . [A]lmost all the experiments on self-perception theory are examples of self-fabrication, whereby people misunderstand the real reason for their behavior and make mistaken inferences about their internal states."[5] This distinction between self-revelation and self-fabrication is at the heart of the difference between genuine revelation and inauthentic, truth-seeming fabrication. And that difference creates what de Man calls irony, something we'd all (most likely) like to avoid in our lives.

To reach some understanding of the difference between fabrication and revelation on our way to self-knowledge requires the comic forces of wit and humor to break down the bastion of rigid, self-conscious rationalism. Kelly Oliver also argues as much when she observes that virtually all contemporary theories of identity reify systems of oppression even as they attempt to break the perpetuating cycles of individuals' dominion over others. Of this "pathology of oppression," Kelly observes that "struggles for recognition and theories that embrace those struggles may indeed presuppose and thereby perpetuate the very hierarchies, domination, and injustice that they attempt to overcome."[6] In other words, attempts to defeat the rationale of oppression by utilizing rationally activated methodologies alone are doomed for the most part to reify the very things they propose to attack. Any system powerful enough to defeat the current system of oppression has built into it the very tools of oppression it claims to overthrow.

Given this point, Oliver draws what we've seen to be the fairly standard comic conclusion. Given that we possess "systems of sensation and perception that operate throughout the entire body" (12) and into the environment at large, then we must accept that "we are fundamentally connected to our environment and other people through the circulation of energies that sustain us" (15), a point that leads logically to assuming an ethical responsibility to respond positively to the erstwhile "othered" world around us, human others certainly included. Oliver's argument is basically in line with Antonio Damasio's argument regarding our environmented entwinement. And it's a significant point to recognize. But Oliver rather incompletely argues for what amounts to be the memic transformation of humanity based primarily on rationally convincing humanity to do the right thing. As we've seen already, however, vulnerabilities abound in this position, given our human urge for individuated self-preservation rather than for species survival. Even if only a solid few chose to cheat in such a world—now dreamily filled to near capacity with gift-giving suckers—the utopia would inevitably crumble.

This is where Wilson's concept of the adaptive unconscious may help with what amounts to comic resolution by unconscious memic transformation. Damasio observes how the nonconscious body necessarily influences consciousness. Damasio and Oliver (and Wilson) would likely agree that it is possible for psychoanalysis to shed light on thoughts and experiences buried in the recesses of the mind, helping consciousness to create a more "complete" vision of selfhood by doing what Oliver sees as the Freudian notion of self-conscious "working-through." However, Wilson argues that there is yet another level of nonconscious being, the "adaptive unconscious," that never sees the light of rationalist or even conscious day. As Wilson notes, the "*adaptive* unconscious is not the same as the *psychoanalytic* one" (5, italics added). Unlike the psychoanalytic unconscious, the adaptive unconscious mechanism "automatically" triggers primitive urges like the four "f's (to fight, flee, feed, or fornicate) that our "civilized" conscious must regulate.

But the adaptive unconscious significantly *also* engages "higher order" functions designed to maximize survivability. Wilson points out that it includes such "higher-order, executive functions as goal setting" (48) as well as "interpretation [and] evaluation" (43) in its pursuit of optimal survivability. For example, while most of us would submit that we are not racists (or sexists or otherwise biased), psychologists have succeeded at proving that our bodies testify otherwise, the result of myriad forces (cultural and otherwise) invading our senses below the radar of our conscious selves. We tense up in the presence of "otherness," react in more defensive or aggressive ways when we encounter unfamiliarity, and so on, all of which undermines our conscious assertions that we are what we "say" we are. Our adaptive unconscious confirms the survivalist selfishness that Dawkins has identified as fundamentally genetic. Such unconscious prejudices abound as a result of our urges to preserve our selves and our kind; and, generally unbeknownst to our conscious selves, they affect our daily behavior in sometimes monumental (and monstrous) ways. For example, humans of European descent "intuitively" prefer to follow leaders who are tall, physically fit, dark and fully haired, white, and male even when such qualities are not necessarily qualities that this leadership position may require. Political beauty contests have frequently resulted in leaders lacking anything more than figurehead qualifications. A similar disconnect holds for such enigmatic phenomena as "love" and "hate," whose various qualities of attraction and repulsion physically hold us in their gravitational fields and defy rational definition, invariably flying in the face of our consciously delineated preferences.[7] Why, exactly, do we love or hate those people we do? Hamlet, as well as most of humanity, would like an answer to that question. Can a catalog of attributes explain why we love who we do? Or why we hate?

When we have this curious and uncomfortable disconnect between our conscious beliefs and explanations and unconscious feelings and behaviors, given this curiously split personality inhering in one body, the best that our consciousnesses can do is invent Wilde-like narratives or fictions that seem reasonably to approach a conscious explanation of our nonconscious and unconscious behaviors. As Wilson notes, "our conscious selves often do not know the causes of our responses and thus have to confabulate reasons" (99).

Important to Wilson's concept, however, is also the idea that this unconscious is *adaptive*. The unconscious "reads" the body's environment as it makes its various unconscious determinations of how best to survive (or thrive) in a given situation. This point leads to the possibility that an effective way to change its (and our) behavior is to alter its (and our) environment. For example, if one chose to alter one's unconscious responses to racial difference, one could choose to move into and adapt to an interracial environment. Familiarity with and comfort in such an environment would or could effectively alter defensive behaviors triggered by our unconscious, aligning our unconscious reactions with our conscious declarations. But first we must learn to understand what it is that our adaptive unconscious is "telling" us. Laughter at a racist or sexist joke, for example, tells us something negative about ourselves, encouraging a revisioning of ourselves. We often consciously don't want to laugh and actually try to subdue it, but laughter often prevails and, occasionally at least, causes conscious embarrassment.

Our exposures to conditions that cause such eruptions enforce the body's prejudices. But the conscious discomfort generated by such eruptions may also catalyze conscious efforts to alter these unconscious eruptions of prejudice by exposing the body to conditions that could alter the body's behaviors, having it conform to our conscious desires. Citing William James' suggestion, Wilson observes, "[T]he more frequently people perform a behavior, the more habitual and automatic it becomes, requiring little effort or conscious attention. . . . [B]ehavior change often precedes changes in attitude and feelings" (212).

Laughing as a result of a rational incongruity—a professed nonsexist laughing at a sexist joke, for example—reveals something of the broader mechanism of disconnected "being" at work, wherein consciousness and the adaptive unconscious grapple to establish some sort of harmonic integration within the one genetic machine they must both occupy. The laughter of Hobbesian superiority, for example, reveals to us something of our hierarchic prejudices against "inferior" others, which we should either embrace or dismantle. And then there's sympathetic laughter when we laugh along with someone else, especially some other we previously conceived to be so "different." Ideologies of oppression crumble under the resonance of sympathetic laughter. In these cases, ignoring laughter or subduing it are agelastic options every bit as much as indulging laughter generated by racism and sexism is agelastic. But comedy says we should try to identify sympathizing behavior that aligns both or all of these selves into the self we want to be.

So while Oscar Wilde's play may not call for revolution against Victoriana (then or now), the play's ironic insinuation into our bodies' "understanding" of our man-made oppressive environment may succeed at opening adaptively unconscious paths that could lead to altering our responses to the "correct-ness" of certain behavior within an orthodox conception of the world around us. It's the feeling of what it's like to think in harmony with one's body that comedy encourages, inspiring an interplay of order and disorder that allows for visions and revision until somehow we "get it right."

Kit Carson to the Rescue

If one needs to observe the tectonics wherein the adaptive unconscious arises, offers itself as a gift-laden alternative to consciously perceived reality, and then grinds against the monolithic mill of culturally valued commodity, one would do well to recall Clifford Odets' 1935 play *Awake and Sing!*. In the play the family patriarch, Jacob, is alternately obsessed with his Caruso records and a dream longing for communist revolution in his adopted American homeland. He's very much a dreamer from the old world whose time has passed in the hustle and bustle of this vibrant, new capitalist world. As the play's directions indicate, he's a "*sentimental idealist with no power to turn the ideal to action.*"[8] While a certain liberal-leaning sympathy may direct an audi-ence to listen to this man as the "voice of the play," it really is a voice that only speaks to that strain of self-conscious species activism that has proven to be so ineffective. Giving Jacob a soap-box opportunity to preach his revolution-ary thoughts would have been a feasible option here had Odets been one to believe in the kind of liberal rationalist transformation Oliver describes.

However, Odets does not pursue that option. Rather he has this defeated idealist decide to convert himself to a commodity by committing suicide and cashing himself in as an insurance policy in order to generate what Jacob hopes to be a nest egg for his grandson Ralph. But what looks at first to be a crushing capitulation to oppressive capitalism transforms in the play into something altogether different. Unbeknownst to Jacob, he has already passed onto Ralph something that he himself only barely possessed: an adaptively unconscious aching for a world of integrated equilibrium that nowhere currently exists. Jacob's little world of Caruso and communism have given Ralph a bodily feel for what could be, a gift of himself that far outvalues the cashed-in life-insurance policy. In the play's closing moments,

Ralph defiantly declares, "Let mom have the dough. I'm twenty-two and kickin'! I'll get along. Did Jake [Jacob] die for us to fight about nickels? No! 'Awake and Sing,' he said. Right here he stood and said it. The night he died, I saw it like a thunderbolt! I saw he was dead and I was born! I swear to God, I'm one week old! I want the whole city to hear it—fresh blood, arms. We got 'em. We're glad we're living" (93).

For Ralph, nothing *materially* has changed by play's end. He rejects his grandfather's reduction to market-value, insurance-policy commodity and accepts the greater gift that Jacob's wistfully generous environmenting has infused into Ralph's unconscious feel for the world. It's the kind of feeling that Ralph wants to perpetuate for himself and generate for those around him, sensing as he does that only in such a transformed world can he find the harmony that he seeks. Of course, recalling for us Oscar Wilde's famous observation that a cynic is "a man who knows the price of everything, and the value of nothing," Ralph's declaration of rebirth may add up to continued poverty in this commodifying world of capital exchange, but the intangible wealth of the gift of Jacob's being invigorates Ralph in ways the money never could. As "*RALPH stands full and strong in the doorway*" (93), we get the feeling that all will be well, against all our consciously aware cynicism that nothing *really* has changed. Ralph chooses not to play the game of the marketplace; rather, he chooses to find a way, perhaps impossibly, to change the game to align with the *feeling* of generosity and humanity he experienced in the presence of his grandfather. But, then, here we are again at the point where Ralph's singular decision will likely never lead to anything but an isolated haven of little real impact on the world at large.

The same kind of determined, adaptive innocence infects William Saroyan's 1939 play, *The Time of Your Life*, written at the outbreak of World War II and during the final days of the Great Depression (though no one yet knew that). To recall yet another of Oscar Wilde's aphorisms, it would be appropriate for an agelast to remind us that "a sentimentalist . . . is a man who sees an absurd value in everything and doesn't know the market price of any single thing" (373).[9] Like *Awake and Sing!*, *The Time of Your Life* is faint, fragile stuff in the eyes of the cynical economic marketplace. But the value it valorizes goes beyond the minting of coins and even beyond the decided sentimentality it generates.

The cast is a varied mix of simple and "unimportant" human beings. Nearly embalmed in hopelessness, the barroom setting of the play—Nick's, "An American Place,"[10] in San Francisco—nonetheless moves to a level of genuine hope against the palpable oppression permeating it. The text's description of Willie "the marble-game maniac" captures the fragile defiance that exists in virtually all of the play's characters. He "*[s]tands straight and pious before the contest. Himself vs. the machine. Willie vs. Destiny. His skill and daring vs. the cunning and trickery of the novelty industry of America, and the whole challenging world. He is the last of the American pioneers, with nothing more to fight but the machine, with no other reward than lights going on and off, and six nickels for one"* (389). Neither Willie nor Saroyan knew at the time the play was produced that real men and women much like this character Willie, steeped in a generous comic vision of faith, would literally from 1941 to 1945 become the force that would prevent the world from being overrun by agelastic fascism. World War II is historical testament to a viscerally generated optimism that looks to material resources to serve a higher end, something that reaches beyond accumulation of wealth and transforms that bounty into gifts of giving sacrifice and altruism, defying the basic urges of individuated survivalism. This is comedy played out on the historical stage, and it goes far to challenge the cynic's critical charge that staged comedy is sentimental drivel.

The mix in this play of simple men and women struggling against "the machine" summarizes the larger comic impulse, struggling to preserve its fragile fabric as the more abrasive world of agelasm grinds against it. Melodramatic inserts flood the play, including Tom and Kitty, two lovers who will wind up married thanks to Tom getting a job and Kitty giving up a desperate flesh-for-profit life of prostitution. Amid the menagerie is yet another rather sentimental character, Harry, who claims to be a natural comedian and tries his stuff out on the bar's denizens, presenting the following monologue to those who will listen:

> George the Greek takes the cue. Chalks it. Studies the table. Touches the cue-ball delicately. Tick. What happens? He makes the three-ball. What do I do? I get confused. *I go out and buy a morning paper.* What the hell do I want with a morning paper? What I *want* is a cup of coffee, and a good used car. I go out and buy a morning paper. Thursday, the twelfth. Maybe the headline's about *me.* I take a quick look. *No. The headline is not about me.* It's about Hitler. Seven thousand miles away. I'm here. Who the hell is Hitler? Who's behind the eight-ball? I turn around. *Everybody's behind the eight-ball!* (428)

No one laughs at the routine, but everyone comforts Harry that it's funny material. What is comic about the story or routine is its subtextual indebtedness to the adaptive unconscious. Harry's innocent narrative creates for his on- and off-stage audiences an environment that taps into the roots of comic commitment by following unconscious instincts only minimally burdened by any conscious barricades. His is a story of concern that begins with his own problems but expands to include everyone's problem. Never reaching the surface of full articulation, Harry's observations tell a story through mood as much as anything else, leaving it to the body's adaptive unconscious to respond as it may, depending on the barriers allowed to stand in its way by consciousness proper. Harry allows his instinctively cooperative inclinations to come close to conscious light thanks to his innocent simplicity, and most of the characters throughout the play are similarly tuned in to be receptive to what he offers. Ultimately, though unconsciously, the headline *is* about him. And his barroom buddies know the headline is about them, too, even if they only know without precisely *knowing* it.

But these vignettes of hope amid quiet desperation are secondary to the apparently overriding self-consciously controlled impulses of Joe, the choreographer of the goings-on in the bar. A success in every business undertaking he's ever attempted, he has all the luck, all the tools, and all the money, it appears, to "fix" everyone's problems. Despite having all the ingredients to become a first-class agelast, Joe is a genuinely sympathetic consciousness who has withdrawn from the world of industry and is now quietly striving to find a better place for himself in hopes of making the world a better place in the bargain "If anybody's got any money—to hoard or to throw away—you can be sure he stole it from other people. Not from rich people who can spare it, but from poor people who can't. . . . I'm no exception. . . . I stole it like everybody else does. I hurt people to get it. Loafing around this way, I *still* earn money. The money itself earns *more*" (457). This wealth, in turn, has driven him to a sense of personal guilt: "I've got a Christian conscience in a world that's got no conscience at all. The world's trying to get some sort of a *social* conscience, but it's having a devil of a time trying to do *that*" (457).

But despite all his best intentions and despite his actual efforts, when it truly matters, Joe fails to complete his mission of saving his acquaintances. This capitalist-turned-benefactor works minor miracles throughout the play, but in the end he is incapable of preventing evil from entering this honky-tonk, which arrives in the form of Blick, a menacing bully of a policeman determined to abuse his authority over this cooperative of apparently power-less innocents.

The man who finally destroys the menace to all in this bar (and into the streets) is the superannuated vagabond calling himself "Kit Carson," a man full of tall tales that utterly defy credulity. If Harry's comedy is poignant and subtle, Kit's humor is full of incongruities and improbabilities and is down-right hilarious. For example (joining in at the middle of a tale):

> KIT CARSON: If I told you that old Southern gentleman was my grandfather, you wouldn't believe me, would you?
> JOE: I might.

KIT CARSON: Well, it so happens he wasn't. Would have been romantic if he had been, though.

JOE: Where did you herd cattle on a bicycle?

KIT CARSON: Toledo, Ohio, 1918.

JOE: Toledo, Ohio? They don't herd cattle in Toledo.

KIT CARSON: They don't anymore. They did in 1918. One fellow did, least-ways. Bookkeeper named Sam Gold. Straight from the Eastside, New York. Sombrero, lariats, Bull Durham, two head of cattle and two bicycles. Called his place The Gold Bar Ranch, two acres, just outside the city limits. That was the year of the War, you'll remember.

JOE: Yeah, I remember, but how about herding them two cows on a bicycle? How'd you do it?

KIT CARSON: Easiest thing in the world. Rode no hands. Had to, otherwise couldn't lasso the cows. Worked for Sam Gold till the cows ran away. Bicycles scared them. They went into Toledo. Never saw hide nor hair of them again. (436–37)

Viewed from an orthodox position of common sense, Kit's incredible tales mark him as little more than a good drinking buddy with virtually no credibility and absolutely no power.

As the play reaches its crisis and Blick's brutality becomes intolerable, Joe tries to fix the problem by killing Blick, taking desperate action only to have his gun fail to fire. Depressed, Joe sinks into his bar chair, only to hear, nearly immediately, the incredible news that Blick has been killed just outside the bar, and no one (not even the cops) is looking for the murderer. Who was it? Kit Carson enters the bar, bloodied from an earlier beating at the hands of Blick, and begins a new tale: "I shot a man once. In San Francisco. Shot him two times. In 1939, I think it was. In October. Fellow named Blick or Glick or something like that. . . . Saw him walking and let him have it, two times. Had to throw the beautiful revolver into the Bay" (480). Joe calmly gets up and walks out the door, looking at Kit "*with great admiration and affection*" (480). Clearly, this world of down-and-out denizens can stand up for itself, presumably even without the intrusion of repentant capitalism. The "American routine" in the bar resumes.

In *The Time of Your Life*, Saroyan has very nearly returned us to the folk wisdom of medieval carnival, implying that the "American spirit" that lives in the common man can hold its own against agelastic tyranny and oppression. The Kit Carson of this play is a parody of the adaptive incarnation of the American myth of self-reliance prior to being co-opted by the urges of self-interest. It's a self-reliance that serves the community at large because in an act of personal self-preservation, Carson actually serves the community into which he's stumbled. There's a faith here that, left to its own devices, the adaptive unconscious can self-organize and self-police its environment when conscious conceit is properly subdued.

But Can Anything *Really* Beat Dominant Culture?

Susan Carlson's 1991 study, *Women and Comedy*, opens with an analysis of Somerset Maugham's 1926 play, *The Constant Wife*. Carlson quotes from Constance, the play's central character whose adulterous husband she appears willing to forgive. However, it's not the standard forgiveness of a wronged but saintly wife that occurs: "[W]hat is a wife in our class? Her house is managed by servants, nurses look after her children, if she has resigned herself to having any, and as soon as they are old enough she packs them off to school. Let us face it, she is no more than the mistress of a man of whose desire she has taken advantage of to insist on a legal ceremony that will prevent him from discarding her when his desire has ceased."[11] Constance decides in the end to go on holiday with another man but promises to remain John's "constant wife," thereby suggesting a new culturally "orthodox" paradigm that overturns the old double standard of male-female behavior while remaining at least formally within the bounds of accepted dominant-culture custom. But, then, a paradigm of isolated celibacy seems (literally) a rather unproductive option.

Using this play and others as support, Carlson summarizes a fundamental problem with comedy and women: "Women's equality to men is not . . . comedy's essential ingredient. Exactly the opposite is true. The fact that comedy's women are *never* quite equal is the key to its traditional definitions of joy, community, and hope" (33). Comedies resulting in happy endings as a result of marriage place women in unequal roles by dominant-culture standards, given the paranoid cultural strictures placed on matrimony by patrimonial authority in general. And these material facts correctly disturb feminist critics of comedy. Furthermore, comedy has no real defense against this criticism; and though there is an explanation, it may nevertheless be quite unsatisfying.

Carlson correctly sees women's liberties becoming formally restricted by comic resolutions ending in marriage. One could also argue, however, that in comedy both women *and* men ultimately accept fewer liberties by play's end than they singly had at play's beginning. But if the comic agenda has succeeded, these early liberties—like autonomy and independence—are no longer desirable given that the comic enterprise has transformed these former egoists of individuated consciousness into people filled with a newly adapted awareness of the mutual benefits of matrimonial cooperation.

A key point here is that men and women arrive at this mutually cooperative destination by separate paths. Recall Judith Butler's observation that the cornerstone of agelastic order is based on a Socratic paradigm that values masculinized ideas over feminized materiality, the very paradigm that comedy challenges. Culturally empowered masculinity, in other words, is the source of agelastic order and control against which comedy must struggle; and the ennatured feminine is the comic force that must work against the pathology of oppression that the masculine ethos endorses. From this perspective, it is possible to see that the feminist and comic agendas can or should intersect at crucial points in their assaults on individuating, masculine, agelastic dominion and all that follows.

If this is an accurate assessment, then the charge of women is to adopt qualities of guarded altruism, the cooperative strategy best suited to reverse the dominating influences of masculine strategies of selfish, individualizing defection from the larger human fold. The altruistic "weapon" of choice, according the Oliver, should be love, the only effective means by which "we can overcome [the] domination" (217) that exists in any fundamentally "master-slave" relationship. Somehow, the dominator must be convinced or willingly "coerced" to give up control, which can only genuinely be overcome through authentic exchanges of love. And somehow women must take the risk of altruistic exposure advocated by the comic agenda. But very importantly, it must be emphasized that altruistic behavior be undertaken not naïvely but somehow guardedly and with eyes wide open.

If the strategy works, then guarded, savvy, and resilient altruists like Portia, Beatrice, and Kate will transform and conform potential tyrants like Bassanio, Benedick, and Petruchio, reminding us that conforming the already empowered may be the secret to the whole problem facing *anyone* who is disenfranchised, not just women. The internal strength of character that such comically heroic action demands is, as we've seen, nearly superhuman, certainly of a sort greater than even proud Hector could muster. When it works, though, it *really* works. Significantly, comedy argues that the strategy we witness the women above using while conforming their erstwhile oppressors is the sort that converts from within rather than, say, at the point of a sword. And the staying power that this transformative brand of conversion generates will be of an order greater than armed coercion could ever expect. Comedy then further extends the hope that such a strategy will pave the way to larger and larger sanctuaries like Belmont, mustering in turn enough critical mass to infiltrate agelastic worlds like Venice with hopes of having accrued enough power to tilt the defectors' balance toward cooperative transformation.

After all, far less savory is the option of the *masculinized* feminine transformations that Caryl Churchill presents in her 1982 play *Top Girls*, reflecting a "feminist" victory co-opted by Thatcherite conservatism resulting in a mere transfer of the patriarchal power ethic to a patriarchally programmed woman. A strategy of lovelessness merely results in a redistribution of power and reinscription of a pathology of oppression.

Recall, most importantly, that with this altruistic strategy must come a hard-nosed pragmatism not always presumed to inhere in comedy. Bernard Shaw's *Major Barbara* (1905) provides a good visionary case in point. Barbara's Salvation Army efforts to convert her society's oppressed criminal element fail miserably, generating only self-interested cheater's hypocrisy among her charges out of her naïvely well-intended sanctimony. Little does she realize that her wayward efforts at transformation are actually culturally endorsed and even institutionalized within the larger community, because she fails fully to understand that she is part of a bad-faith industrial-strength scheme to manipulate her "gift" of salvation in order to promote, prolong, and continue universal subjection, subservience, and suffering. Barbara's utopian plan to generate universal salvation falls victim to a sucker's fate, suffering at the hands of a cheater's marketplace that reduces her human charges to units of labor and uses Barbara's promise of salvation to minimize revolution or at least minimize eruptions of union strikes.

Capitalism in its single-minded support of the individual has done what misguided readings of Darwinism have also done: presumed that self-preservation is best served through autonomous acts of self-interest. It has separated itself from Adam Smith's famous observation that economic self-interest should operate under a providential "invisible hand" of mutual beneficiality. When the providential hand is pushed out of the self-interested marketplace equation, the result is that the charitable "hand" of goodwill must pick up as best it can. Charity and community concerns are left to operate in only limited and peripheral arenas like poverty management and control of revolution. Governments once charged with controlling runaway individualism have instead been manipulated to endorse such behavior and to disrupt cooperative-leaning enterprises that should emulate natural "providence." Barbara's Salvation Army, in effect, is a mollifying, satellite extension of a system that has allowed moral and ethical behavior to function in out-of-the-way corners of society while the mainstream uncaringly surges forward. Her efforts will never effectively enter the dominant circle of commodity exchange and therefore will never be more than a sucker's strategy susceptible to agelastic manipulation. Barbara's actions, finally, actually support the status quo.

Barbara's father, Undershaft, on the other hand, has learned a different and even counterintuitive path to salvation: "[Y]ou must first acquire money enough for a decent life, and power enough to be your own master."[12] For Shaw, wealth and power inoculate against agelast tyranny and oppression, a not necessarily revolutionary thought. In fact it is actually something of a capitalist's idealistic mantra. The trick, though, is neither to hoard riches for purely personal gain, though personal gain may be the natural and initial inspiration. Rather, Undershaft's pursuit of "money and gunpowder" creates the conditions to perpetuate a community of plenty that permits a possessionless sentiment of more humane exchange inherent in a gift rather than commodity market. Wealth and power are to be used for the creation and enhancement of a powerful community that can grow into a real challenge to the dominant cultural status quo. The "invisible hand" has been freed from deceptive agelastic shackles, for we see that rather than vaingloriously accepting credit for what the Undershaft firm has accomplished, Undershaft freely confesses that what drives the whole operation is "A will of which I am a part" (79). And his own wealth and comfort depend completely on the wealth and comfort of those who "work" for him.

Undershaft has, in essence, tapped into his sympathy-leaning adaptive unconscious, which in his case is relatively easy to do given that he is unpolluted by formal education and socialization, being, as he is, a foundling. Unlike his cultured son (raised by mom), Undershaft's unencrusted—or at least less encrusted—adaptive unconscious has more direct access to his untutored decision-making processes, which allows him to feel the beneficial harmony generated by perpetuating an environment of mutual gain and protection. The very idea that *only* a foundling can inherit the Undershaft dynasty is itself acknowledgement and endorsement of the power of the adaptive unconscious.

It takes a while for the well-intended but overeducated and ego-ridden Barbara to see the wisdom her father embodies. But Barbara ultimately accepts the error of her well-intended ways and joins her father, declaring, "My father shall never throw it in my teeth again that my converts were bribed with bread. I have got rid of the bribe of heaven. Let God's work be done for its own sake: the work he had to create us to do because it cannot be done except by living men and women. When I die, let him be in my debt, not I in his; and let me forgive him as becomes a woman of my rank" (89). Decidedly unsentimental and with an edge unexpressed by the likes of Grandpa in *You Can't Take It With You*, Barbara has realized that her Salvation Army brand of salvation was little more than agelast-preserving, sugar-coated bribery feeding her own individualistic self-satisfaction. It was commodity exchange disguised as gift/grace/salvation morality. Though still interested in a balance-sheet ethic, a now guardedly altruistic Barbara is determined that the bottom line will have her owed rather than owing, a sort of potlatch engagement wherein she's determined to give until the culturally empowered recipient ultimately crumbles in the face of overwhelming charity that it can no longer restrain, trick, or cheat.

In a rather post-Victorian declaration of Victorian confidence, Shaw seems to be saying in *Major Barbara* (and elsewhere) that the Darwinian theory of survivability does not merely advocate selfish fitness of the individual so much as it does individual fitness supported by empowering cooperation that triumphs at a larger level when universal benefits achieve critical mass.

Facing Down the Ruling Class

Peter Barnes' 1968 *The Ruling Class* reminds us of the dangers attendant on adopting a comic agenda but failing properly to put up your guard. Even more dangerous would be any decision that entails giving up on the comic altogether. Total victory may never completely belong to comedy, and that is surely a frustrating thought to accept. But it would be utter suicide to give in to a system bent on strangling the chaotic creativity of our adaptive unconscious and abandoning it completely in favor of a controlling, closed consciousness. Beginning with the point that our divided selves reflect and even initiate this struggle at the larger social level, Barnes' play more than amply demonstrates the folly of comic capitulation and giving in to agelasm.

As the play opens, we see the Thirteenth Earl of Gurney giving his unorthodox personal inclinations free reign in the privacy of his chambers, until a chair falls out from under him, and he hangs himself—a grim beginning. But then the closeted behavior we witness in the Thirteenth Earl blasts completely into the full light of day in the form of the Fourteenth Earl of Gurney—Jack. To say he is less inhibited than his father is an understatement. In fact, Jack is quite literally the most uninhibited—and most all-powerful—creature in the cosmos, describing himself as "[t]he creator and ruler of the Universe, Khoda, the One Supreme Being and Infinite Personal Being, Yaweh, Shangri-Ti and El, the First Immovable Mover, Yeah, I am the Absolute Unknowable Righteous Eternal, the Lord of Hosts, the King of Kings, Lord of Lords, the Father, Son and Holy Ghost, the one True God, the God of Love, the Naz!"[13] Any audience member possessing even a slight degree of sanity would conclude that Jack is delusional, especially given the backdrop of stern sobriety that is Jack's powerfully aristocratic family and circle of acquaintances. However, his behavior verifies that this conscious statement of persona or identity aligns in many ways with Jack's free-wheeling adaptive unconscious: he is a genuinely benevolent, loving, blissfully benign individual, dispossessed of any sense of greed or inherited authority, reliant purely on his giving nature. He is, in short, the personification of the creative force of life itself, though perhaps he overstates the case a bit.

His virtues, of course, are not appreciated by the agelastic establishment, bent as it is on maintaining strict control over the world that it claims to possess by hereditary right. No Prince Hal or Portia among its ranks, this ruling class concludes that this personification of the comic spirit cannot be allowed to insinuate its way into the dominant social fabric. The family must protect its power, interests, possessions, and reputation. On the advice of a psychologist that Jack be given a good dose of reality, Jack is introduced to Mr. McKyle, whose life has led him toward a parallel delusional regarding godhead. Using the logic that "it's impossible for two objects to occupy the same space at the same time" (685), Dr. Herder believes Jack will be forced to concede he's not god if he confronts someone with a similar claim. McKyle's vision of god, however, is the direct opposite of Jack's: not a god of love and light but rather "Jehovah o' the Old Testament, the Vengeful God" (687). McKyle's dark vision is built on his experience within his environment: "No god of love made this world. I've seen a girl of four's nails torn out by her father. I've seen the mountains of gold teeth and hair and the millions boiled down for soap" (690). Losing his grip on his own unconscious sense (or faith) in a benevolent god, Jack can only respond, "S-S-some-times G-G-god turns his b-b-back on his p-p-people" (690).

Vulnerably exposed to this onslaught—in part because he has no critical-mass community for protective support—Jack appears to have accepted several unfortunate "facts," most crucially that the world around us proves that god must be other than a god of love and therefore that Jack must be someone other than the god of love. The horrors of the twentieth century can certainty lead to the logical conclusion that any belief in a god of gift and love is delusional since love cannot *logically* coexist in an environment so painfully cruel. It's equally logical to conclude that pursuing a non-zero-sum, godlike/comic agenda of love and gifting will only make us even more vulnerable to the oppressive horrors of a world spiraling downward to inevitable entropic oblivion. Therefore, the way of the world—the way to survive in this world—is to operate in the commodified world of zero-sum measure for measure, hoarding what we can for our selfishly individual salvation, using strict reciprocity as a shield to guard against domination by others, and abandoning the vulnerability and risk necessary for the exercise of a "gift" economy.

Jack converts, brutally abandoning his sense of love and grace that seemed to have welled up from his unconscious inclinations. He accepts the logically self-conscious conclusion that meting out strict, severe, orderly justice is the only way to clean up a world so mired in death, suffering, and decay. Jack eventually converts to a vengeful, remorseless Jack-the-Ripper, a seemingly inevitable end, and he murders his one convert to love, Claire, whose murder in turn capitally condemns his one fellow "revolutionary," Tucker (the Thirteenth Earl's former butler), the man that the ruling class charges with the crime in order to cover the transgressions of the new leader of the ruling class, the newly installed Fourteenth Earl of Gurney.

Barnes manages an ending whose message is indirectly made by all comic material: following a prevailing logic of allowing socially sanctioned order to triumph unchallenged is, quite literally, a dead end.

This point reminds us that comedy is always vulnerable to dark and twisted forces that perpetually appear on the verge of utter triumph. Comedy itself has even been put to dark and twisted use under the agelastically deceptive guise of serving the comic end when such uses in fact do nothing of the kind. This is a point made by Trevor Griffiths' *Comedians* (1976). Griffiths confronts the world of stand-up comics and the practice of eliciting laughter with aggressive jokes directed *against* others, tapping into sentiments of fear and hatred of others and thereby reinforcing a cultural pathology of oppression. Like McKyle in *The Ruling Class*, Eddie Waters (the comic tutor for an evening class on "doing" comedy) has been affected by the horrors of inhumanity inflicted on "others." However, for Eddie the horrors he witnesses operate differently on his consciousness.

Visiting a German concentration camp after the war, Eddie reports that he saw "a special block, 'Der Straf-Bloc,' 'Punishment Block.' It took a minute to register, I almost laughed, it seemed so ludicrous. Then I saw it. It was a world like any other. It was the logic of our world . . . extended. . . . And I discovered . . . there were no jokes left. Every joke was a little pellet, a final solution."[14] The absurdity of a punishment block embedded within a punishment camp resonates for Waters as a sort of larger cultural metaphor. For Waters, the Holocaust fully exposed the extreme consequences of generating order by oppressing the other. And he realizes that the horrors of the Holocaust began with countless minor acts of oppression, perhaps the most innocent-seeming of all being the brand of aggressive humor generated at the expense of others and found popularly consumed in comedy clubs. Allowing for such an environment of minor oppression to fester and breed leads in the extreme to this curiously embedded concentration-camp-within-a-concentration-camp ethic, where torment within torment continues virtually without any sign of ending. Humor turned into a weapon of oppression is the ultimate betrayal of the comic agenda.

As a result, Waters tries to get his students to generate humor that doesn't denigrate "Negroes. Cripples. Defectives. The mad. Women. . . . Workers. [The] Dirty. . . . Unschooled. . . . [and] Shifty" (19). Waters tries to teach his students to avoid the easy laugh of assault and humiliation. But the results of his teachings are mixed at best. Most of his students succumb on stage to the easy trick of inducing audience laughter by targeting precisely who and what Waters argues against. The implication is clear: audiences as consumers of humor, environmented in a world of aggression, almost intuitively prefer the insulting and denigrating brand of aggressive humor to its more unifying, cooperative alternatives. We hunger for confirmation of our superiority in this most Hobbesian of zero-sum cultural circumstances, isolating others as beneath us and therefore worthy of economic misuse—and worse.

Raised in isolation from "others" in our generally segregated lives, our adaptive unconsciousnesses learn through environmental homogeneity to suspect difference. Therefore, the laughter it generates is an aggressive endorsement of prejudices we often consciously deride but unconsciously endorse. Here, a poorly environmented adaptive unconsciousness reveals a need to alter its environmental circumstances rather than to reinforce them by encouraging eruptions of hostile laughter. Or, sadly, we can continue to endorse implicit prejudice against others by welcoming such laughter, oppressively and selfishly to the detriment of any communally harmonizing urges we may *memically* claim to have.

One student out of Waters' group uses his skills in a manner that could be described as aggressive, but he takes the energy in an altogether different direction. His routine is complex and difficult exactly to summarize. Andrew Stott notes the result: "Through the disjointed and coarse dialogue, Price conveys the inarticulacy and anger he perceives in the working-class male, marginalized by middle-class society and reduced to expressing himself through violence and sexual aggression."[15] Stott adds, "No conventional description would call it comedy, however" (118), actually reminding us of the young comic in *The Time of Your Life*, whose description of quietly desperate existence likewise evoked a sympathetic but not laughing response. Price's routine inspires and catalyzes unity instead of fomenting fragmentation. This comic "sees" the relationship between the angry and aggressive working-class consciousness and the environment its alienated unconscious is forced to experience. As a result, his routine resists humiliating workers, revealing instead an implicit longing for homeostatic balance that can't be achieved in the culturally oppressive world he and they occupy. This new comic looks for points of unifying sympathy, finding the raw material that is at its simplest the heart of comedy and the greatest threat to dominant culture.

Lessons Beyond Tolerance: A Comic Resolve

There is little doubt that Tony Kushner's widespread mainstream success with *Angels in America, Part One* (1991) and *Part Two* (1992)[16] has made him a voice to be reckoned with. His political activism would at first glance seem to preclude his being capable of creating transformative comedy, committed as he appears to be to affirming, politicized confrontation. But in *Angels in America*, he arguably succeeds at the transformative over and above any affirming, activist designs he may have intended, following Brecht, it seems, beyond the political and truly into the comic.

The satiric political target is an agelast authority epitomized by American Reaganomics of the 1980s, arguably the high point of American self-confidence. In *Angels in America, Part One*, Joe, the gay Mormon lawyer married to the pill-popping Harper, describes the Reagan era in terms that would make Republican pundits proud even today: "I think things are starting to change in the world. . . . Change for the good. America has rediscovered itself. Its sacred position among nations. And people aren't ashamed of that like they used to be. This is a great thing. The truth restored. Law restored. That's what President Reagan's done. . . . He says 'Truth exists and can be spoken proudly.' And the country responds to him. We become better. More good" (26). These are words of the Republican Right easily relevant even today in the post-9/11 era: truth exists; we possess the truth; we will "share" our truth with those who are currently benighted. Actually, if we distil out allusion to Reaganism, Joe's speech speaks to the general longings of most Americans, Republican and otherwise. It encapsulates an "American" optimism that is in many ways this general America's biggest export, marketed and sold throughout the globe.

However, what this optimism sells runs the risk of providing the opposite of what it claims, as is evidenced in the life of Louis, who in the play is something of the typical pro-American American. Louis describes himself as a person who believes that the world "will change for the better with struggle." He "has this neo-Hegelian positivist sense of historical progress towards happiness or perfection or something . . . [and] feels very powerful because he feels connected to these forces, moving uphill all the time" (25). Pretty standard American optimism. But Louis' comfy vision is confronted and undermined by the counter-progressive reality of AIDS, made painfully real by the infection of his lover, Prior.

Louis is not a bad man. He's certainly not a Republican. What he is is a man who has bought into an illusionary vision of reality as a result of being an *American*. But this seemingly benign "comic" American dream of unfettered optimism is challenged by the realities of AIDS, which reveal that the illusion of harmonizing "American" unity is a brittle thing easily fragmented when placed under stress. Under the pressure of the growing AIDS epidemic, the unity of the social circle is disrupted as individuals run for survivalist cover.

Kushner further problematizes this optimism by presenting American culture as ossifying into a bleak bureaucratic wasteland dressed up as paradise, afraid to (or incapable of) change despite the overwhelming output of frenetic heat, energy, and paperwork. Much energy is expended, but little progress actually occurs. After all, the heaven we aspire to create on earth is little more than a safe, secure, homogeneous, white-bread world imitating death in all ways except that we remain breathing amid the static world of sameness. It's a world that increasingly leads us to what, in *A Bright Room Called Day*, Kushner describes as "A DIM AND OPPRESSIVE AWARENESS AMONG THE PEOPLE THAT THE BATTLE HAS TURNED AWAY FROM THE STREETS AND THE BALLOT BOX TO SECRET DEALS BETWEEN POWERFUL PEOPLE IN PRIVATE ROOMS."[17] That is Kushner's description of Fascist Movement advances in Germany in 1931. Insert Halliburton, Rumsfeldt, Chaney, Bush, and big business, however, and Kushner would likely argue that the description aptly applies to a post-9/11 America.

Something very telling occurs in Louis' "Why has democracy succeeded in America" speech (*Part Two*, 89–90), where he argues that most Americans don't understand freedom: "[L]ike they see these bourgeois property-based Rights-of-Man-type rights but . . . that's just liberalism, the worst kind of liberalism, really, bourgeois tolerance, and what I think is that what AIDS shows us is the limits of tolerance, that it's not enough to be tolerated, because when the shit hits the fan you find out how much tolerance is worth" (90). Amid all of what occurs in *Angels in America*, this brief section perhaps most significantly articulates the comic, transformational agenda of the play, touching as it does on the concept of tolerance. Tolerance is an operation in which one entity—the empowered—accepts the presence of an other, and it is a recognition that can be withdrawn at any point that it may be in the better interest of the empowered to do so. It's an agelastically deceptive enterprise generating a false illusion of cooperation and community. If it can be withdrawn, it is certainly not a "gift," and if it entails obligation and even debt, then it cannot be conceived to be a way to eliminate our culture's pathology of oppression.

Whether it's the gay and lesbian community, cultures of the Middle East or the former Soviet bloc (all subjects of Kushner's opus), or any other Other that doesn't quite conform to white American heterosexual male Protestant orthodoxy, Kushner refines yet one more cornerstone point about comedy: tolerance alone is not an equitable end. In 1985, Dorothy Riddle developed a scale known as "Homophobic Levels of Attitude,"[18] which echoes Louis' (and Kushner's) sentiment revealing exactly how much higher than tolerance we need to aim. The scale identifies a continuum beginning with "repulsion" and ending with "nurturance." In Kushner's scheme, orthodox culture is simply *repulsed* by anything that fails to conform to its ideal, the most extreme position on Riddle's scale. At this level, the belief is that anything is justified to change nonconformist behavior: prison, hospitalization, even violence. "Remove others from our presence" is the agenda. We're clearly in the world of extreme agelasm, of fascist intolerance. At the next level, "pity," we see a position that presumes that the Other is inferior and that conversion to dominant behavior should be enforced: "Make others better by making them like us." This is at least one explanation for the motive behind Shylock's enforced conversion, one that most current *Merchant of Venice* audiences cannot allow to pass without at least implicit conscious discomfort.

Just above pity comes "tolerance," seemingly the goal of all well-intended citizens. We talk tolerance, and leaders preach tolerance. As Kushner notes, however, when situations strain the quality of tolerance by triggering self-preservationist urges among the powerful, an otherwise transparent and generous sense of indulgence turns to impatience and often worse. Our aggressive individualism rises, and self-survival takes the center stage of our consciousness. Tolerate a child's crying, until it gets on our nerves, that is. Tolerate Muslims or a gay lifestyle until their rights challenge our privilege. Tolerance is thin protection and dissolves when strained. Tolerance is not equity or justice. Consider the success of America's "tolerant" separate but equal racial doctrine. Tolerance clearly is not sufficient.

The next reported step is "acceptance," but even this level places those who are "accepted" in the vulnerable position of requiring protection against an ever-present possibility of a dominant culture's withdrawal of support. It is actually the step beyond "acceptance" where we begin to move into a world of equity and justice as espoused by affirming political activism as well as comic transformation. This level is what Louis wants, though he doesn't articulate the word. And it is Kushner's aim as well. "Support" is the first proper level, where working to safeguard rights is central. Moving to "admiration," "appreciation," and "nurturance"—levels that actually value diversity and otherness as something that enhances quality of life for all—are levels we are far from reaching as a culture. These are what Kushner's theatre strives to achieve

Louis' speech makes it clear that Kushner believes tolerance is not enough. But Kushner goes a step further by generating in the play a full experience of what Louis articulates. Insisting as he does in presenting a range of flawed, terribly human, unapologetically gay men on stage is exactly the thing to do to test the resilience of "tolerance" among his play's audiences. Does our tolerance allow us to watch the play without feeling our bodies unconsciously flinching just a little at the play's sometimes graphic depiction of otherness behavior? Do we respond as we do to a racist joke, feeling our bodies react against the other even as our consciousness strives to sympathize with the other? Kushner tests his audience, at least in the early stages of the play, in exactly that way. For those of us with a "merely" tolerant attitude, using a weakly conscious posture to hide an unconscious prejudice, the play should be a humbling lesson on just what tolerance means.

Paving the way for a change in unconscious predispositions is what comedy does. Educating the adaptive unconsciousnesses of the consciously predisposed is a distinctly *realistic* comic goal, leading to a deep understanding that tolerance alone is not sufficient, that maybe legal partnership is not "as good as" marriage, that maybe sitting down with those "benighted" Muslims and Palestinians in the Middle East would be better than creating police states to control their "unruly" behavior until we can democratize them. And maybe doing such things will benefit both our individual self-preservationist urges *and* the self-preservationist longings of others, to the mutual benefit of all and detriment of none.

Kushner ends his epic *Angels* with a motley community of damaged but mutually giving protagonists sitting around the winterized fountain of Bethesda (in New York's Central Park), a symbol of restored health, awaiting its springtime resuscitation. Prior ends the play by announcing to the audience,

> You are fabulous creatures, each and every one.
> And I bless you: *More Life.*
> The Great Work Begins. (*Part Two*, 148)

Life is nothing without comic hope, especially in the darkness before dawn, during the frost that precedes the thaw.

When Utopia?

Where and when exactly will a world of mutually beneficial equity and justice prevail over individual self-interest? We probably should be reminded that the word "utopia" literally translates to "no where." As such, perhaps the absolute best we can ever really hope for is a fleeting glimpse of that world. To draw on the title of Stoppard's 2003 three-play epic, *The Coast of Utopia*, perhaps we are forever destined to cruise offshore against a headwind that prevents landing, running the risk of shipwreck and annihilation as we struggle for a better look at what can't be reached.

The play proper takes us through the mid-1800s, primarily following a band of Russian intellectual revolutionaries initially swept away by ideological fervor for freedom imported from Germany and France. The long history that Stoppard plays out leads to the shipwreck of the 1848 French Revolution, which demonstrated that the pathology of oppression is actually self-imposed by victims of oppression as much as by their oppressors. Though the spirit of revolution kept alive by these idealists, leading to the Russian emancipation of the serfs in 1861, there is a pervading sense of failure generated by comparing accomplishments against their initial goals. The high ideals of complete justice and equity were far from achieved. But perhaps finally there's more accomplishment here than initially meets the eye.

In the trilogy's final installment, *Salvage*, Karl Marx announces his vision about human existence, that "[e]very stage leads to a higher stage in the permanent conflict which is the march of history."[19] Once "Capital and Labour stand revealed in fatal contradiction" says Marx, "[t]hen will come the final titanic struggle, the last turn of the great wheel of progress beneath which generations of toiling masses perished for the ultimate victory. Now at last the unity and rationality of history's purpose will be clear to everyone. . . . Everything that seemed vicious, mean, and ugly, the broken lives and ignoble deaths of millions, will be understood as part of a higher reality, a superior morality, against which resistance is irrational—a cosmos where every atom has been striving for the goal of human self-realisation and the culmination of history" (118–19). Deluded by his own idealism, Marx has clearly lost contact with any sense of a "natural" inclination toward individuated self-preservation. What he fails to realize is that simply trying to will away, or even trying to reason away, individuated self-preservationist urges is a foolish enterprise. Seeing a grand species march toward rationally inevitable, linearly certain salvation, Marx is clearly lost in the clouds of Sokrates' Thinkery.

Alexander Herzen, having fully experienced a life of love, sacrifice, and death, contradicts Marx's fully mechanistic, utterly nonhuman, and nearly agelastic vision: "But history has no culmination. . . . History knocks at a thousand gates at every moment, and the gatekeeper is chance. We shout into the mist for this one or that one to be opened for us, but through every gate there are a thousand more. We need wit and courage to make our way while our way is making us. But that is our dignity as human beings, and we rob ourselves if we pardon us by the absolution of historical necessity" (119). The nearly infinite choices that humanity has before it reflects the chaotic freedom that looms before us all. Self-conscious choice and determined action are humanity's charge, and it's not so much a collective charge as a charge placed before every individual within the species. But, then, at the same time we are also unavoidably part of the unexpected turns of history that are beyond our control. And so we have options to consider and pursue.

At one extreme is the anarchist, Michael Bakunin, who Falstaff-like utterly indulges his urges with a lust that stands out in the work as unadulterated *joie de vivre*, even (or especially) as he suffers incarceration and exile. Through it all, Bakunin offers a boisterous summation of existence: "Seven degrees of human happiness! First, to die fighting for liberty; second, love and friendship; third, art and science; fourth a cigarette; fifth, drinking; sixth, eating; and seventh, sleeping" (117). Bakunin reveals that living, finally, is the attraction that Marx has left out of his formula. Though Marxism is frequently held up as an ideal to be pursued, it is just that: an ideal that, if pursued, can only really guarantee "the ignoble deaths of millions."

But fighting for liberty should not involve some far-away culmination; rather, as Herzen argues, "A distant end is not an end but a trap. The end we work for must be closer, the labourer's wage, the pleasure in the work done, the summer lightning of personal happiness" (119). If Bakunin is one extreme and Marx another, Herzen himself has finally come to a sort of middle place on the scale, though by play's end he moves much closer to Bakunin than to Marx. Quietly, nearing exhaustion but not resignation, Bakunin observes, "We have to open men's eyes and not tear them out . . . and if we see differently, it's all right, we don't have to kill the myopic in our myopia. . . . We have to bring what's good along with us" (119–20).

Given the aggressive individualism that is our genetic lot, the end result of utopia that Marx envisions is not a likelihood; rather moments of communal harmony and social justice will erupt as individuals choose one door rather than another and enter one kind of relationship over another. Waiting for the end result of actual, full utopia is an idealism every bit as dangerous as any agelastic ideal previously dreamed up by human consciousness. Living now but also working for change is the comic charge.

Stoppard's 2006 work, *Rock 'n' Roll*, makes much the same point, moving from 1968 to 1990 and following the Czech resistance to communist oppression leading to the Velvet Revolution. The Falstaffian existence in the play takes on two forms, one being the Piper, who later turns out to be Syd Barrett, the irrepressible heart of the band Pink Floyd. The other is the Czech band, Plastic People of the Universe, bent as they are on playing rock 'n' roll without any regard to political consciousness or its like. As the play unfolds, what becomes evident is that Barrett and the Plastic People become manifestations of the spirit of emancipation, refusing cooption by either capitalist profit-making ventures or by political systems trying to manipulate their fame. The danger here for agelasm is that such spirits of comedy are uncontrollable because they are uncorruptible. They have no interest in being absorbed into a commodities economy of zero-sum competition and aggression. And they inspire activism among others by creating a unifying outsider's environment of selfless self-determination, generating a force that assumes, eventually, a critical mass against which agelasm itself must capitulate. On the surface, of course, political activism gets the credit. But a closer look may reveal that such activism may never have eventuated had it not been for the spirit of comedy behind the movement.

But then, once again we are reminded of the painful shortcomings and incompleteness that attend pursuing the utopian dream. Much like the French Revolution of 1848, this Velvet Revolution that is so central to *Rock 'n' Roll* does not fulfill its utopian potential so much as it paves the way for the pursuit of individualized creature comforts in Eastern Europe. Admittedly for the Czech Republic in 1990, the character Jan is right to say, "These are new times. Who will be rich? Who will be famous?"[20] However, it's obviously not a time of utopian optimism, merely a time for capitalist and individualist self-indulgence. But can such a move be completely unexpected, given that "humanity" is really not a group noun but a term identifying a vast body of individuals naturally scurrying for personal survival?

On those occasions when cooperation benefits a body of individuals, then the semblance of species interest will appear. The goal of the comic agenda is to encourage pursuing such occasions as frequently as possible as we move through the chance circumstances that make up our lives. Self-interest is with us; wishing it to be otherwise is the heart of idealistic folly. So says comedy.

CONCLUSION

Things Could Be Better, Things Could Be Worse

Lear goes mad on the heath. Macbeth has a seizure at a banquet. Othello has his own seizure. And Hamlet struggles mightily to keep his sanity. These conscious wills in conflict with their unconscious bodies show us the results of excluding comic influences from our existences in favor of abstract disembodied idealisms. These bodily eruptions against their agelastic oppressors reveal untapped undercurrents between our environmented bodies and our willful, headstrong desires to disembody and autonomize our minds. The result in the cases above is tragedy.

If chaos theory is correct and if nature in its orderly disorderliness is neither really malign nor benign in its rhythms, then attending to its irregularly patterned and unpredictable laws is not necessarily a bad thing, and there is no crime in accepting the nature that is "us." Comedy reminds us of our connection to nature by reminding us of our bodilyness. And, in turn, comedy reminds us that attention must be paid to the body's adaptive unconscious. This doesn't mean that rational, orderly, even abstracting consciousness has no place in our lives. Pure carnival indulgence, unadulterated celebration of the body's urges, is not comedy's final end, lest we become little more than Calibans searching for our Mirandas in order to complete our singular genetic destinies merely to people the world with more of our kind. Rather, comedy reminds us that consciousness alone is insufficient—that, unchecked by the body's adaptive unconscious, consciousness can lead to disequilibrium and separation from our vital connections to our bodies and the world it and therefore we inhabit. Recall where we'd be if rationality were our only crutch: logically speaking, we would be intellectually mired in a stultifying worldview paralleling Zeno's paradoxes, a world amazingly seduced by dead-end philosophies that logically, rationally, and intellectually reveal the insubstantiability of virtually everything we once held dear and encouraging little more than debilitating guilt leading to Prufrock-like indecision and inaction.

Or when we do take up arms in the name of an ideal untested by the body's scrutiny, we end up with catastrophe. Tom Stoppard reminds us of this in *The Coast of Utopia*, dramatizing the European shipwrecks of 1848. He further notes that the 1861 emancipation of the serfs (much like America's emancipation of it slaves) meant little when the ideal of freedom was unattached to a way to sustain the body. In the play, Natalie Ogarev makes the point: "The peasants are told they're free, and they think the land they've worked now belongs to them, even the big house belongs to them, and the livestock and probably Madame's Paris frocks, too—so when it turns out nothing belongs to them and they have to pay rent for their plots, well, obviously, freedom bears an uncanny resemblance to serfdom."[1] The body tells us that we need to eat and that no amount of "ideal" can sustain us for long, defeating that rationalist/intellectual gravitational force that insists we pursue a higher level of existence without first considering our physical sustainability.

Comedy's unique gift is that it reveals that powerful but shadowy transforming force, the adaptive unconscious, in ways that virtually no other human enterprise can. Political satire or psychoanalytic analyses may connect shadows of (un)consciousness to conscious awareness and develop some sense of a need for change. It may even present viable alternatives, utopian or pragmatic. But so can so many other forms of affirmingly conscious expression. Comedy, however, is the rare transformative tool that allows our unconsciousnesses to "talk" to our consciousnesses, our bodies to our minds. In the best of circumstances, comedy encourages the two to talk and in the process arrive at consensus.

As many have noted, comedy does often appear to provide an unfortunate endorsement of the status quo. And it often toys with radical subversion and revolutionary thought only to return to a traditional, unchanged, unchanging order, typically through traditional marriage or reinstitution of previous political or economic orders. But that generic reinscription called a happy ending is often merely a conscious convention that, in many instances, should be downplayed in favor of the fundamental open-endedness expressed prior to the artifice of play's closure. That celebration of open-endedness springs from our body's unconscious and is the heart of the comic spirit itself. Comedy creates an environment that can encourage our consciousnesses to comprehend and adapt to the wisdom enigmatically expressed by our unconsciousnesses. Through comic exposure—and perhaps unawares—we become more understanding of the other we've previously excluded, whether it is previously *othered* ideas, peoples, or cultures. Perhaps an egalitarian and even altruistic sentiment will enter the conscious via unconscious infection. Perhaps we will actually become as "understanding" as our consciousnesses frequently (though often inaccurately) claim we are. And perhaps this process will lead us to loosen hold on patterns of patriarchal domination and measure-for-measure reciprocity, which would actually feed back into those suspect happy endings given that they would then embody something other than what they have traditionally been conceived to be.

Stoppard once again summarizes how this transformation may best be able to occur. Alexander Herzen summarizes, "I'm beginning to understand the trick of freedom. Freedom can't be a residue of what was unfreely given up, divided up like a fought-over loaf. Every giving up has to be self-willed, freely chosen, unenforceable. Each of us must forgo only what we choose to forgo, balancing our personal freedom of action against our need for the cooperation of other people—who are each making the same balance for themselves."[2] He then echoes what is a central comic insight insofar as the world is currently constituted: "What is the largest number of individuals who can pull off this trick? I would say that it's smaller than a nation, smaller than the ideal communities of Cabet or Fourier. I would say the largest number is smaller than three. Two is possible, if there is love, but two is not a guarantee" (66). In small units—starting off as small as units of two—we must learn the value of giving up certain rights and entitlements in order to see forward into a non-zero-sum world that can ultimately provide even greater benefits than we currently singularly strive to accrue. From such humble beginnings, however, grows the dream of growing such units into actual communities, into freely generated confederations of individuals benefiting from and grown stronger by numbers, making them collectively more and more capable of confronting and restraining agelastic aggression and oppression.

Politics can legislate behavioral change, but it can't really do more than force a sort of ungifted tolerance on a consciousness whose bodily unconscious remains resistant. Psychoanalysis can uncover the recessed emotions and memories repressed by our overbearing conscious awareness. But the unconscious that is often the domain of psychoanalysis is little more than the shadowy recesses of consciousness itself, for it is really nothing like the body's *adaptive* unconscious. For better or for worse, this curious and enigmatic adaptive force operates independently and with virtually no connection in either direction to our conscious will. We can consciously at best glimpse traces of the trajectory of the adaptive unconsciousness, in the form of nervous twitches, reflex reactions, and laughter. But then we have a curious tool to play with those traces, testing the validity of what they seem to be telling us, or working to change our body's behavior should it be telling us something unsavory about our selves. The tool is comedy. Losing this unique access, or denying its import, we all run the risk of becoming Lears, Macbeths, Othellos, and Hamlets, struggling against our bodies' better inclinations and heading us into our own shipwrecks off the many fabled coasts of utopia.

Many current theorists of comedy have announced the death of comedy or, at best, its reduction to a dark and glowering modification of its former self. Andrew Stotts observes in his 2005 analysis of comedy, "One of the less traumatic but nevertheless noticeable effects of the events of 9/11 was the voluntary moratorium on humour that immediately followed it in the American media."[3] Few would argue against the appropriateness of such a moratorium on a temporary, respectful basis. But total abandonment of humor—and abandonment of virtually all aspects of the comic agenda—in this post-Holocaust, post-9/11 world would be far more than merely inappropriate because it is precisely in times of greatest crisis that comic attunement to our adaptive unconscious is indispensable. If in previous generations there was an outcry against "vulgar," trivial comedy and the time wasted indulging such "lower" urges, it would be well for this and future generations to realize that, without the comic urge in our midst, humanity puts itself in great jeopardy. Uncritically or even "heroically" abandoning the hope of comic conciliation results in demonizing others in ways that build chasms rather than bridges, creating devastating Peloponnesian Wars rather than negotiated détente.

This is not to say that we should uncritically embrace comic appeasement, declaring peace in our time even as we open the gates to cataclysm. Game theory demonstrates that conciliatory cooperation preached by comedy clearly has its place. But we would do well to at least listen to cynical doomsayers and accept the idea that altruism must muster the strength actually to be able to defend itself against agelastic oppression. Fully unguarded extensions of free love and gifts of other sorts in the face of dominant-culture deception and treachery is just so much foolishness, verging on an idealism itself worthy of tragic commentary. However, to announce the death of comedy altogether is a pronouncement indeed premature—as always has been the case throughout comedy's long life.

Comedy has always provided hope even (especially) during our darkest days. Shortly after hurricanes Katrina and Rita devastated New Orleans and much of its gulf-coast surroundings, my hometown of Baton Rouge became a city-sized M.A.S.H. unit, witnessing a mass influx of newly homeless and apparently broken victims. Amid the palpable despair that one would expect from such a catastrophe, however, there abounded among the victims a quiet optimism that frankly had no rational justification. Somehow, fresh from rescue and still stewing in the septic stench of the only clothes they had, these neighbors from sixty miles away could still smile, joke, and even laugh as they told horrific stories of survival. The living bodies of these men and women somehow sustained their spirits, reminding them that they are tangibly *of* New Orleans, a place tapped into the comic spirit like no other. This spirit will certainly serve the city well during the coming years of rebuilding. The city that care forgot has more grit, more embodied resilience than can be rationally explained. Fully in the best of ways, New Orleans personifies the carnivalesque spirit of comedy that empowers hope over despair. Mardi Gras has always been more than a scheme by New Orleans to rake in tourist dollars. Despite the commodified manifestations of Mardi Gras that this city sells to the rest of the world, that annual bacchanal has given New Orleans much more than New Orleanians perhaps fully realized and certainly more than they ever felt they would need.

If comedy can bring light as we endure life's darker hours, it also has the capacity to anticipate and mobilize against what may loom ahead. And no darker hours loom ahead of humanity than the growing threat of an ecological disaster motivated by individuated self-interest. However, increasingly attuned today to what comedy reminds us is our inseparable connection to our environment, we seem on the verge of shaking off the Platonic idealism that has led us to conceive of ourselves as masters of our domain free to use it as we choose. The inconvenient truth is that our struggle to master all the others of the universe has exacerbated rather than minimized the threats to our *individual* existences by jeopardizing the very environment without which we cannot live. Of late, the ecological movement has succeeded at forcing agelastic forces to concede human complicity in global warming, thanks to a critical mass of cooperating individuals patiently converting individuated resistance to sympathetic cooperation. The threat to individual quality existence has succeeded at reaching at least a promise of cooperation large enough to enact change for the good of all.

At this crucial stage, it must be remembered that cooperative postures in general provide perfect conditions for individuating cheaters to thrive. The degree to which the guarded altruists of this current movement can prevent a cheater's backlash against this cooperative enterprise depends on the comic resilience of these very real comic heroes in our midst. Recall that the best strategy involves the generosity of gift, which includes transformatively convincing rather than affirmatively coercing the recipient to accept what is offered. Punitive alternatives *may* force behavioral alterations, and such changes *may* alter adaptive inclinations. But the resistance will likely remain as palpable as Shylock's resistance to Christian conversion. As Oliver notes, conversion through the bonding element of love is very likely the best way for humanity to save itself—and unite its countless individuated selves. In the end, the results will determine whether Samuel Beckett's denatured plays of bleak entropy were visionary or cautionary tales. Let's hope the latter.

There will likely never be publicly proclaimed accounts of comedy changing the world, being the behind-the-scenes operator that it is. But without comedy, it is less likely that we will ever have (or would ever have had) the disposition to enact changes or to endure hardships of any real import. Comedy is the chaos that disrupts orthodox stasis and negativity, and comedy undermines the tyranny of rigid consciousness. It goes without saying that the world has lots of room for improvement. Less obvious is the point that without comedy the world would very likely not have come along as far as it has.

Notes

Introduction

1. Sean O'Casey, *Juno and the Paycock*, in *Three Plays*, Sean O'Casey (London: Macmillan, 1957), 18.
2. Joseph W. Meeker, *The Comedy of Survival: Literary Ecology and a Play Ethic*, 3rd ed. (Tuscon: University of Arizona Press, 1997), 12.
3. Kirby Olson, *Comedy after Postmodernism* (Lubbock: Texas Tech University Press, 2001), 6. Hereafter cited in the text.
4. Philip Auslander, *Presence and Resistance: Postmodernism and Cultural Politics in Contemporary American Performance* (Ann Arbor: University of Michigan Press, 1992), 23.
5. Judith Butler, *Bodies that Matter: On the Discursive Limits of "Sex"* (New York: Routledge, 1993), 13, 10. Hereafter cited in the text.
6. Kelly Oliver, *Witnessing: Beyond Recognition* (Minneapolis: University of Minnesota Press, 2001), 3.
7. William E. Gruber, *Comic Theaters: Studies in Performance and Audience Response* (Athens: University of Georgia Press, 1986), 4. Hereafter cited in the text.
8. Erich Segal, *The Death of Comedy* (Cambridge, MA: Harvard University Press, 2001).

Chapter 1

1. Aristotle, *Poetics*, in *Aristotle's Poetics. Translation and Analysis*, Kenneth A. Telford (Chicago: Henry Regnery, 1961), 10. Hereafter cited in the text.
2. For a tantalizing suggestion of what Aristotle's treatise on comedy might have included, see Umberto Ecco's imaginative, fragmentary re-creation in *The Name of the Rose*, trans. William Weaver (New York: Harcourt, 1984), 468.
3. Henry Alonzo Myers, *Tragedy: A View of Life* (Ithaca, NY: Cornell University Press, 1956), 45. Quoted in Nathan A. Scott, Jr., "The Bias of Comedy and the Narrow Escape into Faith," *Christian Scholar* 44, no. 1 (Spring 1961): 17.
4. Scott, "Bias of Comedy," 17. Hereafter cited in the text.
5. Terry Eageleton, *Sweet Violence: The Idea of the Tragic* (Oxford: Blackwell, 2003), 43–44. Hereafter cited in the text.

6. Jan Walsh Hokenson, *The Idea of Comedy: History, Theory, Critique* (Madison, NJ: Fairleigh Dickinson University Press, 2006).

7. Hokenson notes that it was Robert M. Torrance in 1978 who first identified the distinction between the satiric and populist/celebratory threads in comedy. See Robert M. Torrance, *The Comic Hero* (Cambridge, MA: Harvard University Press, 1978).

8. Jerry Aline Flieger, *The Purloined Punchline: Freud's Comic Theory and the Postmodern Text* (Baltimore: Johns Hopkins University Press, 1991), 258–59. Hereafter cited in the text.

9. Susan Purdie, *Comedy: The Mastery of Discourse* (Toronto: University of Toronto Press, 1993), 8. Hereafter cited in the text.

10. See Joseph Meeker, *The Comedy of Survival: Literary Ecology and a Play Ethic*, 1974 (Tuscon: University of Arizona Press, 1997); and Robert Storey, "Comedy, Its Theorist, and the Evolutionary Perspective," *Criticism* 38, no. 3 (Summer 1996), 407–41. Hereafter cited in the text.

11. Henri Bergson, *Laughter: An Essay on the Meaning of the Comic*, trans. Cloudesley Brereton and Fred Rothwell (New York: Macmillan, 1917), 10. Hereafter cited in the text.

12. Quoted in Robert Storey, "Comedy, Its Theorists, and the Evolutionary Perspective," *Criticism* 38, no. 3 (Summer 1996): 407. Hereafter cited in the text.

13. See Richard Dawkins, *The Selfish Gene*, 1976 (Oxford: Oxford University Press, 1989), 189–201.

14. See Nancy Fraser, *Justice Interruptus* (New York: Routledge, 1997), 23, 24.

15. Kelly Oliverm Witnessing: Beyond Recognition (Minneapolis: University of Minnesota Press, 2001), 50. Hereafter cited in the text.

16. Jessica Blank and Erik Jensen, *The Exonerated* (New York: Faber and Faber, 2004), 75.

17. J. L. Styan, *The Dark Comedy: The Development of Modern Comic Tragedy*, 2nd ed. (Cambridge: Cambridge University Press, 1968), 297.

18. Erich Segal, *The Death of Comedy* (Cambridge, MA: Harvard University Press, 2001), 452. Hereafter cited in the text.

19. Aldous Huxley, "Tragedy and the Whole Truth," in *A Book of English Essays*, selected by W. E. Williams, 265–66 (Harmondsworth, UK: Penguin Books, 1948). Quoted in Scott, "Bias of Comedy," 21.

20. Herbert Blau, "Comedy since the Absurd," in *The Eye of Prey: Subversions of the Postmodern* (Bloomington: Indiana University Press, 1987), 18.

21. Teresa Brennan, *The Transmission of Affect* (Ithaca, NY: Cornell University Press, 2004), 24.

Chapter 2

1. Wylie Sypher, "The Meanings of Comedy," in *Comedy: Meaning and Form*, ed. Robert W. Corrigan (San Francisco: Chandler, 1965), 18.

2. Susanne Langer, "The Comic Rhythm," in *Comedy: Meaning and Form*, ed., Robert W. Corrigan (San Francisco: Chandler, 1965), 126.

3. Erich Segal, *The Death of Comedy* (Cambridge, MA: Harvard University Press, 2001), 12. Hereafter cited in the text.

4. Émile Durkheim, *The Elementary Forms of Religious Life*, 1912 (New York: Free Press, 1995). For a discussion of Baudrillard's debt to Durkheim, see William Merrin, *Baudrillard and the Media: A Critical Introduction* (Cambridge, England: Polity, 2005).

5. Jean Baudrillard, "Objects, Images, and The Possibility of Aesthetic Illusion," in *Jean Baudrillard: Art and Artefact*, ed. Nicholas Zurbrugg (London: Sage, 1997), 11.

6. Jean Baudrillard, *The Ecstasy of Communication* (New York: Semiotext(e), 1988), 75.

7. Jean Baudrillard, *Seduction* (London: Macmillan, 1990), 144.

8. Joseph W. Meeker, *The Comedy of Survival: Literary Ecology and a Play Ethic*, 3rd ed. (Tuscon: University of Arizona Press, 1997), 37. Hereafter cited in the text.

9. Antonio Damasio, *Descartes' Error: Emotion, Reason, and the Human Brain* (New York: G. P. Putnam, 1994), 90. Hereafter cited in the text.

10. Antonio Damasio, *Looking for Spinoza: Joy, Sorrow, and the Feeling Brain* (Orlando, FL: Harcourt, 2003), 12.Hereafter cited in the text.

11. Antonio Damasio, *The Feeling of What Happens: Body and Emotion in the Making of Consciousness* (Orlando, FL: Harcourt Brace, 1999), 49.

Chapter 3

1. Antonio Damasio, *Descartes' Error: Emotion, Reason, and the Human Brain* (New York: G. P. Putnam, 1994), 225. Hereafter cited in the text.

2. See Jonas Barish, *The Antitheatrical Prejudice* (Berkeley: University of California Press, 1981).

3. Mary Thomas Crane, *Shakespeare's Brain: Reading with Cognitive Theory* (Princeton, NJ: Princeton University Press, 2001), 14.

4. David Hillman, "Visceral Knowledge: Shakespeare, Skepticism, and the Interior of Early Modern Realism," in *Body in Parts: Fantasies of Corporeality in Early Modern Europe*, ed. Hillman and Carla Mazzio (New York: Routledge, 1997), 83.

5. E. M. W. Tillyard, *The Elizabethan World Picture*, 1943 (Harmondsworth, UK: Penguin, 1972), 78–79.

6. Una Ellis-Fermor, *The Jacobean Drama, An Interpretation* (New York: Vintage, 1964), 99.

7. Helen Ostovich, "Introduction," in Ben Johnson, *Every Man Out of His Humour* (Manchester: Manchester University Press, 2001), 14.

8. Ben Johnson, *Every Man Out of His Humour*, 1600 (Manchester: Manchester University Press, 2001), Ind. 123. Hereafter cited in the text.

9. Richard Wilbur, "Introduction," in Moliere *The Misanthrope and Tartuffe*, trans. Richard Wilbur, 1954 (New York: Harcourt, Brace, and World, 1965), 7.

10. Henri Bergson, *Laughter: An Essay on the Meaning of the Comic*, trans. Cloudesley Brereton and Fred Rothwell (New York: Macmillan, 1917), 138.

11. Susan Purdie, *Comedy: The Mastery of Discourse* (Toronto: University of Toronto Press, 1993), 7.

12. See Johan Huizinga, *Homo Ludens: A Study of the Play Element in Culture*, 1944 (Boston: Beacon, 1950).

13. R. D. V. Glasgow, *Madness, Masks, and Laughter: An Essay on Comedy* (Madison, NJ: Fairleigh Dickinson University Press, 1995), 167. Hereafter cited in the text.

14. W. H. Auden, "Notes on the Comic," in *Comedy: Meaning and Form*, ed. Robert W. Corrigan (San Francisco: Chandler, 1965), 66–67.

15. Wylie Sypher, "The Meanings of Comedy," in *Comedy: Meaning and Form*, ed. Robert W. Corrigan (San Francisco: Chandler, 1965), 50.

16. Erich Segal, *The Death of Comedy* (Cambridge, MA: Harvard University Press, 2001), 16.

17. Mikhail Bakhtin, *Rabelais and His World*, 1936, trans. Helene Iswolksky (Cambridge, MA: MIT Press, 1968), 141. Hereafter cited in the text.

18. L. E. Pinsky, "The Laughter of Rabelais," in Mikhail Bakhtin, *Rabelais and His World*, 1936, trans. Helene Iswolksky (Cambridge, MA: MIT Press, 1968), 141.

19. Suzanne Langer, *Feeling and Form: A Theory of Art* (New York: Charles Scribner's and Sons, 1953), 344.

20. Nathan A. Scott, Jr., "The Bias of Comedy and the Narrow Escape into Faith," *The Christian Scholar* 44, no. 1 (Spring 1961): 27.

21. George Meredith, "On the Idea of Comedy, and of the Uses of the Comic Spirit," in *George Meredith's* Essay On Comedy *and Other* New Quarterly Magazine *Publications*, ed. Maura C. Ives (Lewisburg, PA: Bucknell University Press, 1999), 140. Hereafter cited in the text.

22. Antonin Artaud, "The Theatre and the Plague," in *The Theatre and Its Double*, 1938, trans. Mary Caroline Richards (New York: Grove, 1958), 31.

23. John Gay, *The Beggar's Opera*, 1728, in *British Dramatists from Dryden to Sheridan*, ed. George H. Nettleton, Arthur E. Case, and George Winchester Stone, Jr. (Carbondale: Southern Illinois University Press, 1969), 530–65.

24. Wylie Sypher, "The Meanings of Comedy," in *Comedy: Meaning and Form*, ed. Robert W. Corrigan (San Francisco: Chandler, 1965), 23.

Chapter 4

1. Robert M. Torrance, *The Comic Hero* (Cambridge, MA: Harvard University Press, 1978), 11.

2. Simon Critchley, *On Humour* (London: Routledge, 2002), 51.

3. Berthold Brecht, *The Threepenny Opera*, in *Brecht The Threepenny Opera, Baal, The Mother*, trans. Ralph Manheim and John Willett (New York: Arcade, 1993), 126–27. Hereafter cited in the text.

4. Antonio R. Damasio, *Looking for Spinoza: Joy, Sorrow, and the Feeling Brain* (New York: Harcourt, 2003), 275–76. Hereafter cited in the text.

5. Antonio R. Damasio, *Descartes' Error: Emotion, Reason, and the Human Brain* (New York: G. P. Putnam's Sons, 1994), 123.

6. Daniel C. Dennett, *Consciousness Explained* (Boston: Little, Brown, 1991), 23.

7. Damasio, *Looking for Spinoza*, 28.

8. Judith Butler, *Bodies that Matter: On the Discursive Limits of "Sex"* (New York: Routledge, 1993), 48.

9. Erich Segal, *The Death of Comedy* (Cambridge, MA: Harvard University Press, 2001), 44. Hereafter cited in the text.

10. Aristophanes, *Lysistrata*, in *Four Plays by Aristophanes*, trans. William Arrowsmith, Richmond Lattimore, and Douglass Parker (New York: Meridian, 1994), 350. Hereafter cited in the text.

11. See Jacques Derrida, *The Gift of Death*, trans. David Wills (Chicago: University of Chicago Press, 1995).

12. William Arrowsmith, "Introduction," in *Four Plays by Aristophanes*, trans. William Arrowsmith, Richmond Lattimore, and Douglass Parker (New York: Meridian, 1994), 17.

13. Aristophanes, *The Clouds*, in *Four Plays by Aristophanes*, trans. William Arrowsmith, Richmond Lattimore, and Douglass Parker (New York: Meridian, 1994), 25. Hereafter cited in the text.

14. See, for example, Aristophanes, *The Clouds*, in *The Complete Plays of Aristophanes*, ed. Moses Hadas (New York: Bantam, 1962), 102.

15. Aristophanes, *The Acharnians*, in *The Complete Plays of Aristophanes*, ed. Moses Hadas (New York: Bantam, 1962), 27.

16. Kelly Oliver, *Witnessing: Beyond Recognition* (Minneapolis: University of Minnesota Press, 2001), 68.

17. Damasio, *Looking for Spinoza*, 274.

Chapter 5

1. Friedrich Nietzsche, *The Birth of Tragedy* and *The Genealogy of Morals*, trans. Francis Golffing (New York: Doubleday Anchor, 1956), 19. Hereafter cited in the text.

2. Wylie Sypher, "The Meanings of Comedy," in *Comedy: Meaning and Form*, ed. Robert W. Corrigan (San Francisco, Chandler, 1965), 23. Hereafter cited in the text.

3. David Bohm and F. David Peat, *Science, Order, and Creativity* (New York: Bantam, 1987), 127. Hereafter cited in the text.

4. John Fleming, *Stoppard's Theatre: Finding Order in Chaos* (Austin: University of Texas Press, 2001), 70.

5. Tom Stoppard, *Galileo*, quoted in John Fleming, *Stoppard's Theatre: Finding Order in Chaos*, 71.

6. Tom Stoppard, *Arcadia* (London: Faber and Faber, 1993), 48. Hereafter cited in the text.

7. Stephen H. Kellert, *In the Wake of Chaos* (Chicago: University of Chicago Press, 1993), 15–16.

8. See Steven Rose, *Lifelines: Biology beyond Determinism* (New York: Oxford University Press, 1998).

9. See Daniel C. Dennett, *Consciousness Explained* (Boston: Little, Brown, 1991).

10. Tom Stoppard, *Travesties* (New York: Grove, 1975), 62. Hereafter cited in the text.

11. Tom Stoppard, *Artist Descending a Staircase* (London: Faber and Faber, 1988), 19.

12. Tom Stoppard, *Jumpers* (New York: Grove, 1972), 12–13. Hereafter cited in the text.

13. Antonio R. Damasio, *The Feeling of What Happens: Body and Emotion in the Making of Consciousness* (Orlando, FL: Harcourt, 1999), 16.

14. See John Maynard Smith, *Evolution and the Theory of Games* (Cambridge: Cambridge University Press, 1982).

15. Robert Axelrod, *The Evolution of Cooperation* (New York: Basic Books, 1984), 89.

16. See Richard Dawkins, *The Selfish Gene*, 1976 (Oxford: Oxford University Press, 1989). Hereafter cited in the text.

17. Kelly Oliver, *Witnessing: Beyond Recognition* (Minneapolis: University of Minnesota Press, 2001), 224.

18. Tom Stoppard, *The Invention of Love* (London: Faber and Faber, 1997), 43.

Chapter 6

1. Tom Stoppard, *Jumpers* (New York: Grove, 1972), 70

2. Richard Dawkins, *The Selfish Gene*, 1976 (Oxford: Oxford University Press, 1989), 192. Hereafter cited in the text.

3. Jacques Derrida, *Given Time: 1. Counterfeit Money*, trans. Peggy Kamuf (Chicago: University of Chicago Press, 1992), 30. Hereafter cited in the text.

4. See Marcel Mauss, *The Gift: The Forms and Reasons for Exchange in Archaic Societies*, 1925, trans. W. D. Halls (New York: Routledge, 1990). Hereafter cited in the text.

5. Mark Osteen, "Introduction," in *The Question of the Gift: Essays Across Disciplines*, ed. Mark Osteen (New York: Routledge, 2002), 4. Hereafter cited in the text.

6. Andrew Cowell, "The Pleasures and Pains of the Gift," in *The Question of the Gift: Essays Across Disciplines*, ed. Mark Osteen (New York: Routledge, 2002), 291.

7. Lewis Hyde, *The Gift: Imagination and the Erotic Life of Property* (New York: Random House, 1979), 4. Hereafter cited in the text.

8. Pierre Bourdieu, "Marginalia—Some Additional Notes on the Gift," in *The Logic of the Gift: Toward an Ethic of Generosity*, ed. Alan D. Schrift (New York: Routledge, 1997), 234.

9. For more on this notion, see Alain Testart, "Uncertainties of the 'Obligation to Reciprocate': A Critique of Mauss," in *Marcel Mauss: A Centenary Tribute*, ed. Wendy James and N. J. Allen (New York: Berghahn, 1998), 103–4

10. Adam Smith, *An Inquiry into the Nature and Causes of the Wealth of Nations*, ed. R. H. Campbell and A. S. Skinner (Oxford: Clarendon, 1981), vol. II, 175–76, quoted by Eun Kyung Min, "Adam Smith and the debt of gratitude," in *The Question of the Gift*, ed. Mark Osteen, 132.

11. Eun Kyung Min, "Adam Smith and the debt of gratitude," in *The Question of the Gift*, ed. Mark Osteen, 133.

12. David Mamet, *American Buffalo* (New York: Grove, 1976), 93.

13. Mark Osteen, "Gift or Commodity?" in *The Question of the Gift*, ed. Mark Osteen, 233.

14. Ronald A. Sharp, "Gift Exchange and the Economics of Spirit" in *The Merchant of Venice*," *Modern Philology* 83, no. 3 (February 1986): 253. Hereafter cited in the text.

15. See Alain Caillé, "The Double Inconceivability of the Pure Gift," *Angelaki: Journal of the Theoretical Humanities* 6, no. 2 (1993): 287–313.

16. Moss Hart and George S. Kaufman, *You Can't Take It With You* (New York: Dramatists Play Service, 1937), 44. Hereafter cited in the text.

Chapter 7

1. Oscar Wilde, *The Importance of Being Earnest, A Trivial Comedy for Serious People*, 1895 (New York: Avon Books, 1965), 23.

2. Alexander Leggatt, *English Stage Comedy, 1490–1990* (New York: Routledge, 1998), 30.

3. Michael Patrick Gillespie, *The Aesthetics of Chaos. Nonlinear Thinking and Contemporary Literary Criticism* (Gainesville: University Press of Florida, 2003), 105.

4. Paul de Man, "The Rhetoric of Temporality," in Paul de Man, *Blindness and Insight: Essays in the Rhetoric of Contemporary Criticism*, 2nd ed. (London: Routledge, 1983), 212.

5. Timothy D. Wilson, *Strangers to Ourselves. Discovering the Adaptive Unconscious* (Cambridge, MA: Belknap, 2002), 206, 207. Hereafter cited in the text.

6. Kelly Oliver, *Witnessing: Beyond Recognition* (Minneapolis: University of Minnesota Press, 2001), 9. Hereafter cited in the text.

7. In addition to Wilson's book on the subject, Malcolm Gladwell's accessible book, *Blink: The Power of Thinking Without Thinking* (New York: Little, Brown, 2005), is an excellent introduction to this concept.

8. Clifford Odets, *Awake and Sing!*, 1935, in *Famous American Plays of the 1930s*, ed. Harold Clurman (New York: Dell, 1959), 22. Hereafter cited in the text.

9. Oscar Wilde, *Lady Windermere's Fan*, 1892, in *The Writings of Oscar Wilde* (Oxford: Oxford University Press, 1989), 373.

10. William Saroyan, *The Time of Your Life*, 1939, in *Famous American Plays of the 1930s*, ed. Harold Clurman (New York: Dell, 1959), 387. Hereafter cited in the text.

11. W. Somerset Maugham, *The Constant Wife*, 1926, in *Collected Plays of W. Somerset Maugham, vol. 2* (London: Heinemann, 1931) 160, quoted in Susan Carlson, *Women and Comedy: Rewriting the British Theatrical Tradition* (Ann Arbor: University of Michigan Press, 1991), 28. Hereafter cited in the text.

12. George Bernard Shaw, *Major Barbara*, 1923, in *George Bernard Shaw's Plays*, 2nd ed., ed. Sandie Byne (New York: Norton, 2002), 243. Hereafter cited in the text.

13. Peter Barnes, *The Ruling Class*, in *Landmarks of Modern British Drama, vol. 1: The Plays of The Sixties* (London: Methuen, 1985), 643. Hereafter cited in the text.

14. Trevor Griffiths, *Comedians* (London: Faber and Faber, 1979), 64. Hereafter cited in the text.

15. Andrew Stott, *Comedy* (New York: Routledge, 2005), 118. Hereafter cited in the text.

16. Tony Kushner, *Angels in America. A Gay Fantasia on National Themes. Part One: Millennium Approaches* (New York: Theatre Communications Group, 1991), and *Angels in America. A Gay Fantasia on National Themes. Part Two: Perestroika* (New York: Theatre Communications Group, 1992). Hereafter cited in the text.

17. Tony Kushner, *A Bright Room Called Day* (New York: Theatre Communications Group, 1994), 51.

18. The scale is locatable on numerous Web sites, of which the following is one: http://gaylib.com/text/homophobia.htm.

19. Tom Stoppard, *Salvage, The Coast of Utopia, Part III* (New York: Grove, 2002), 118. Hereafter cited in the text.

20. Tom Stoppard, *Rock 'n' Roll* (London: Faber and Faber, 2006), 107.

Conclusion

1. Tom Stoppard, *Salvage, The Coast of Utopia, Part III* (New York: Grove, 2002), 90.

2. Tom Stoppard, *Shipwreck, The Coast of Utopia, Part II* (New York: Grove, 2002), 65–66. Hereafter cited in the text.

3. Andrew Stott, *Comedy* (New York: Routledge, 2005), 103.

Bibliography

Aristophanes. *The Complete Plays of Aristophanes*. Edited by Moses Hadas. New York: Bantam, 1962.

————. *Four Plays by Aristophanes*. Translated by William Arrowsmith, Richmond Lattimore, and Douglass Parker. New York: Meridian, 1994.

Aristotle. *Poetics*. In *Aristotle's Poetics. Translation and Analysis*, edited by Kenneth A. Telford. Chicago: Henry Regnery, 1961.

Artaud, Antonin. *The Theatre and Its Double*. 1938. Translated by Mary Caroline Richards. New York: Grove, 1958.

Auden, W. H. "Notes on the Comic." In *Comedy: Meaning and Form*, edited by Robert W. Corrigan, 61–72. San Francisco: Chandler, 1965.

Auslander, Philip. *Presence and Resistance: Postmodernism and Cultural Politics in Contemporary American Performance*. Ann Arbor: University of Michigan Press, 1992.

Axelrod, Robert. *The Evolution of Cooperation*. New York: Basic Books, 1984.

Bakhtin, Mikhail. *Rabelais and His World*. 1965. Translated by Helene Iswolksky. Cambridge, MA: MIT Press, 1968.

Barish, Jonas. *The Antitheatrical Prejudice*. Berkeley: University of California Press, 1981.

Barnes, Peter. *The Ruling Class*. In *Landmarks of Modern British Drama, vol. 1: The Plays of The Sixties*, 627–732. London: Metheun, 1985.

Baudrillard, Jean. *The Ecstasy of Communication*. New York: Semiotext(e), 1988.

————. "Objects, Images, and the Possibility of Aesthetic Illusion." In *Jean Baudrillard: Art and Artefact*, edited by Nicholas Zurbrugg, 7–18. London: Sage, 1997.

————. *Seduction*. London: Macmillan, 1990.

Beckett, Samuel. *Waiting for Godot*. 1954. New York: Grove, 1986.

Bergson, Henri. *Laughter: An Essay on the Meaning of the Comic*. Translated by Cloudesley Brereton and Fred Rothwell. New York: Macmillan, 1917.

Blank, Jessica, and Erik Jensen. *The Exonerated*. New York: Faber and Faber, 2004.

Blau, Herbert. "Comedy since the Absurd." In *The Eye of Prey: Subversions of the Postmodern*, 14–41. Bloomington: Indiana University Press, 1987.

Bohn, David, and F. David Peat. *Science, Order, and Creativity*. New York: Bantam, 1987.

Bourdieu, Pierre. "Marginalia—Some Additional Notes on the Gift." In *The Logic of the Gift: Toward an Ethic of Generosity*, edited by Alan D. Schrift, 234. New York: Routledge, 1997.

Brecht, Berthold. *The Threepenny Opera*. In *Brecht: The Threepenny Opera, Baal, The Mother*, translated by Ralph Manheim and John Willett, introduction by Hugh Rorrison. New York: Arcade, 1993.

Brennan, Teresa. *The Transmission of Affect*. Ithaca, NY: Cornell University Press, 2004.

Butler, Judith. *Bodies that Matter: On the Discursive Limits of "Sex."* New York: Routledge, 1993.

Caillé, Alain. "The Double Inconceivability of the Pure Gift." *Angelaki: Journal of the Theoretical Humanities* 6, no. 2 (1993): 287–313.

Carlson, Susan L. *Women and Comedy: Rewriting the British Theatrical Tradition*. Ann Arbor: University of Michigan Press, 1991.

Congreve, William. *The Way of the World*. In *British Dramatists from Dryden to Sheridan*. 1700, edited by George H. Nettleton, Arthur E. Case, and Winchester Stone, Jr., 307–47.Carbondale: Southern Illinois University Press, 1969.

Cowell, Andrew. "The Pleasures and Pains of the Gift." In *The Question of the Gift: Essays Across Disciplines*, edited by Mark Osteen, 280–97. New York: Routledge, 2002.

Crane, Mary Thomas. *Skakespeare's Brain: Reading with Cognitive Theory*. Princeton, NJ: Princeton University Press, 2001.

Critchley, Simon. *On Humour*. London: Routledge, 2002.

Damasio, Antonio. *Descartes' Error: Emotion, Reason, and the Human Brain*. New York: G. P. Putnam, 1994.

———. *The Feeling of What Happens: Body and Emotion in the Making of Consciousness*. Orlando, FL: Harcourt Brace, 1999.

———. *Looking for Spinoza: Joy, Sorrow, and the Feeling Brain*. Orlando, FL: Harcourt, 2003.

Dawkins, Richard. *The Selfish Gene*. 1976. Oxford: Oxford University Press, 1989.

de Man, Paul. "The Rhetoric of Temporality." In *Blindness and Insight: Essays in the Rhetoric of Contemporary Criticism*, 2nd ed., edited by Paul de Man, 212. London: Routledge, 1983.

Dennett, Daniel C. *Consciousness Explained*. Boston: Little, Brown, 1991.

Derrida, Jacques. *The Gift of Death*. Translated by David Wills. Chicago: University of Chicago Press, 1995.

———. *Given Time: 1. Counterfeit Money*. Translated by Peggy Kamuf. Chicago: University of Chicago Press, 1992.

Durkheim, Émile. *The Elementary Forms of Religious Life*.1912. New York: Free Press, 1995.

Eagleton, Terry. *Sweet Violence: The Idea of the Tragic*. Oxford: Blackwell, 2003.

Ecco, Umberto. *The Name of the Rose*. Translated by William Weaver. New York: Harcourt, 1984.

Ellis-Fermor, Una. *The Jacobean Drama, An Interpretation*. New York: Vintage, 1964.

Fleming, John. *Stoppard's Theatre: Finding Order in Chaos.* Austin: University of Texas Press, 2001.

Flieger, Jerry Aline. *The Purloined Punchline: Freud's Comic Theory and the Postmodern Text.* Baltimore: The Johns Hopkins University Press, 1991.

Fraser, Nancy. *Justice Interruptus.* New York: Routledge, 1997.

Gay, John. *The Beggar's Opera.* 1728. In *British Dramatists from Dryden to Sheridan,* edited by George H. Nettleton, Arthur E. Case, and Winchester Stone, Jr., 530–65. Carbondale: Southern Illinois University Press, 1969.

Gillespie, Michael Patrick. *The Aesthetics of Chaos. Nonlinear Thinking and Contemporary Literary Criticism.* Gainesville: University Press of Florida, 2003.

Gladwell, Malcolm. *Blink: The Power of Thinking Without Thinking.* New York: Little, Brown, 2005.

Glasgow, R. D. V. *Madness, Masks, and Laughter.* Madison, NJ: Fairleigh Dickinson University Press, 1995.

Gould, Stephen Jay. *Ever Since Darwin: Reflections in Natural History.* New York: Norton, 1977.

Grawe, Paul H. *Comedy in Space, Time, and the Imagination.* Chicago: Nelson-Hall, 1983.

Griffiths, Trevor. *Comedians.* London: Faber and Faber, 1979.

Gruber, William E. *Comic Theaters: Studies in Performance and Audience Response.* Athens: University of Georgia Press, 1986.

Gutwirth, Marcel. *Laughing Matter: An Essay on the Comic.* Ithaca, NY: Cornell University Press, 1993.

Hart, Moss, and George S. Kaufman. *You Can't Take It With You.* New York: Dramatists Play Service, 1937.

Hillman, David. "Visceral Knowledge: Shakespeare, Skepticism, and the Interior of Early Modern Realism." In *Body in Parts: Fantasies of Corporeality in Early Modern Europe,* edited by Hillman and Carla Mazzio, 81–105. New York: Routledge, 1997.

Hokenson, Jan Walsh. *The Idea of Comedy: History, Theory, Critique.* Madison, NJ: Fairleigh Dickinson University Press, 2006.

Horner, Robyn. *Rethinking God as Gift: Marion, Derrida and the Limits of Phenomenology.* New York: Fordham University Press, 2001.

Huizinga, Johan. *Homo Ludens: A Study of the Play Element in Culture.* 1944. Boston: Beacon, 1950.

Huxley, Aldous. "Tragedy and the Whole Truth." In *A Book of English Essays,* selected by W. E. Williams. Harmondsworth, UK: Penguin Books, 1948.

Hyde, Lewis. *The Gift: Imagination and the Erotic Life of Property.* New York: Random House, 1979.

Jonson, Ben. *Three Plays: Volpone, Epicoene, The Alchemist.* New York: Hill and Wang, 1961.

———. *Three Plays, Volume Two: Sejanus, Every Man in His Humour, Bartholomew Fair.* New York: Hill and Wang, 1961.

Kellert, Stephen H. *In the Wake of Chaos.* Chicago: University of Chicago Press, 1993.

Kushner, Tony. *Angels in America. A Gay Fantasia on National Themes. Part One: Millennium Approaches.* New York: Theatre Communications Group, 1991.

———. *Angels in America. A Gay Fantasia on National Themes. Part Two: Perestroika.* New York: Theatre Communications Group, 1992.

———. *A Bright Room Called Day.* New York: Theatre Communications Group, 1994.

Langer, Susanne. "The Comic Rhythm." In *Comedy: Meaning and Form*, edited by Robert W. Corrigan, 119–40. San Francisco: Chandler, 1965.

———. *Feeling and Form: A Theory of Art.* New York: Charles Scribner's Sons, 1953.

Leggatt, Alexander. *English Stage Comedy, 1490–1990.* New York: Routledge, 1998.

Levin, Harry. *Playboys and Killjoys: An Essay on the Theory and Practice of Comedy.* New York: Oxford University Press, 1987.

Mamet, David. *American Buffalo.* New York: Grove, 1976.

Maugham, W. Somerset. *The Constant Wife.* In *Collected Plays of W. Somerset Maugham, vol. 2.* London: Heinemann, 1931.

Mauss, Marcel. *The Gift: The Forms and Reasons for Exchange in Archaic Societies.* Translated by W. D. Halls. New York: Routledge, 1990.

Meeker, Joseph W. *The Comedy of Survival: Literary Ecology and a Play Ethic.* 1974. 3rd ed. Tuscon: University of Arizona Press, 1997.

Meredith, George. "On the Idea of Comedy, and of the Uses of the Comic Spirit." 1877. In *George Meredith's* Essay On Comedy *and Other* New Quarterly Mazazine *Publications*, edited by Maura C. Ives. Lewisburg, PA: Bucknell University Press, 1999.

Merrin, William. *Baudrillard and the Media: A Critical Introduction.* Cambridge, UK: Polity, 2005.

Min, Eun Kyung. "Adam Smith and the Debt of Gratitude." In *The Question of the Gift: Essays Across Disciplines*, edited by Mark Osteen, 132–46. New York: Routledge, 2002.

Molière. *The Misanthrope and Tartuffe.* Translated by Richard Wilbur. New York: Harcourt, Brace, and World, 1965.

Myers, Henry Alonzo. *Tragedy: A View of Life.* Ithaca, NY: Cornell University Press, 1956.

Nietzsche, Friedrich. *The Birth of Tragedy* and *The Genealogy of Morals.* Translated by Francis Golffing. New York: Doubleday Anchor, 1956.

O'Casey, Sean. *Juno and the Paycock.* 1924. In *Three Plays*, 1–73. London: Macmillan, 1957.

Odets, Clifford. *Awake and Sing!* 1935. In *Famous American Plays of the 1930s*, edited by Harold Clurman, 19–93. New York: Dell, 1959.

Oliver, Kelly. *Witnessing: Beyond Recognition.* Minneapolis: University of Minnesota Press, 2001.

Olson, Kirby. *Comedy after Postmodernism.* Lubbock: Texas Tech University Press, 2001.

Osteen, Mark. "Gift or Commodity?" In *The Question of the Gift: Essays Across Disciplines*, edited by Mark Osteen, 229–47. New York: Routledge, 2002.

———. "Introduction." In *The Question of the Gift: Essays Across Disciplines*, edited by Mark Osteen, 1–41. New York: Routledge, 2002.

Ostovich, Helen. "Introduction." In *Every Man Out of His Humour*, Ben Johnson, 1–95. Manchester: Manchester University Press, 2001.

Pinsky, L. E. "The Laughter of Rabelais." In *Rabelais and His World*, 1965, Mikhail Bakhtin, translated by Helene Iswolksky, 141. Cambridge, MA: MIT Press, 1968.

Purdie, Susan. *Comedy: The Mastery of Discourse*. Toronto: University of Toronto Press, 1993.

Rose, Steven. *Lifelines: Biology beyond Determinism*. New York: Oxford University Press, 1998.

Saroyan, William. *The Time of Your Life*. 1939. In *Famous American Plays of the 1930s*, edited by Harold Clurman. New York: Dell, 1959.

Scott, Nathan A., Jr. "The Bias of Comedy and the Narrow Escape into Faith." *Christian Scholar* 44, no. 1 (Spring 1961): 9–39.

Segal, Charles. *Lucretius on Death and Anxiety: Poetry and Philosophy in* De Rerum Natura. Princeton, NJ: Princeton University Press, 1990.

Segal, Erich. *The Death of Comedy*. Cambridge, MA: Harvard University Press, 2001.

Shakespeare, William. *The Complete Pelican Shakespeare*. Edited by Stephen Orgel and A. R. Braunmuller. New York: Penguin, 2002.

Sharp, Ronald A. "Gift Exchange and the Economics of Spirit in *The Merchant of Venice*." *Modern Philology* 83, no. 3 (February 1986): 250–65.

Shaw, Bernard. *George Bernard Shaw's Plays*. 2nd ed. Edited by Sandie Byne. New York: Norton, 2002.

Smith, Adam. *An Inquiry into the Nature and Causes of the Wealth of Nations*. Edited by R. H. Campbell and A. S. Skinner. Oxford: Clarendon, 1981.

Smith, John Maynard. *Evolution and the Theory of Games*. Cambridge: Cambridge University Press, 1982.

Stevens, Scott Manning. "The Sacred Heart and Secular Brain." In *The Body in Parts: Fantasies of Corporeality in Early Modern Europe*, edited by David Hillman and Carla Mazzio, 263–82. New York: Routledge, 1997.

Stoppard, Tom. *Arcadia*. London: Faber and Faber, 1993.

———. *Artist Descending a Staircase*. London: Faber and Faber, 1988.

———. *The Coast of Utopia: Part I (Voyage), Part II (Shipwreck), Part III (Salvage)*. New York: Grove, 2002.

———. *The Invention of Love*. London: Faber and Faber, 1997.

———. *Jumpers*. New York: Grove, 1972.

———. *Rock 'n' Roll*. London: Faber and Faber, 2006.

———. *Travesties*. New York: Grove, 1975.

Storey, Robert. "Comedy, Its Theorists, and the Evolutionary Perspective." *Criticism* 38, no. 3 (Summer 1996): 407–41.

Stott, Andrew. *Comedy*. New York: Routledge, 2005.

Styan, J. L. *The Dark Comedy: The Development of Modern Comic Tragedy*. 2nd ed. Cambridge: Cambridge University Press, 1968.

Sypher, Wylie. "The Meanings of Comedy." In *Comedy: Meaning and Form*, edited by Robert W. Corrigan, 18–60. San Francisco: Chandler, 1965.

Tawney, R. H. *Religion and the Rise of Capitalism*. 1926. New Brunswick, NJ: Transaction, 1998.

Testart, Alain. "Uncertainties of the 'Obligation to Reciprocate': A Critique of Mauss." In *Marcel Mauss: A Centenary Tribute*, edited by Wendy James and N. J. Allen, 103–4. New York: Berghahn, 1998.

Tillyard, E. M. W. *The Elizabethan World Picture*. 1943. Harmondsworth, UK: Penguin, 1972.

Torrance, Robert M. *The Comic Hero*. Cambridge, MA: Harvard University Press, 1978.

Wilbur, Richard. "Introduction." In *The Misanthrope and Tartuffe*, translated by Richard Wilbur, 7–10. 1954. New York: Harcourt, Brace, and World, 1965.

Wilde, Oscar. *The Importance of Being Earnest, A Trivial Comedy for Serious People*. 1895. New York: Avon Books, 1965.

———. *The Writings of Oscar Wilde*. Oxford: Oxford University Press, 1989.

Wilson, Timothy D. *Strangers to Ourselves. Discovering the Adaptive Unconscious*. Cambridge, MA: Belknap, 2002.

Index

acceptance, 174
adaptation, 19, 69
adaptive unconscious, 157–58, 163, 171, 175, 179, 180, 181, 182
admiration, 174
Adorno, Theodor, 26
affirmative remedies, 24, 171, 183
agelasm, 72, 73, 75, 89, 124, 128, 139, 154, 155, 161, 163, 165, 166, 167, 168–69, 170, 174, 177, 181
Alchemist, The (Jonson), 77–80
alchemy, 77–78, 82
altruism, 120–24, 127, 128, 133, 139, 144–45, 148, 153, 161, 165, 181
American Buffalo (Mamet), 134–36
anecomincs, 135, 138
Angels in America (*Part One* and *Part Two*), 171–75
Apollonian, 103–4, 105
Arcadia (Stoppard), 106, 107–12
Archanians, The (Aristophanes), 93–95, 100
Aristophanes, 9, 14, 25, 72, 92–100, 103, 104, 107, 145. *See also specific plays*
Aristotle, 11, 12, 13, 14, 15, 107
Arrowsmith, William, 96
Artaud, Antonin, 77
Auden, W. H., 64, 106
Auslander, Philip, 5–6, 7
authenticity, 44–45, 156
Awake and Sing! (Odets), 159–60
Axelrod, John, 120–21

Bakhtin, Mikhail, 3, 71, 74–76, 85, 88, 93, 139
Barish, Jonas, 55
Barnes, Peter, 168, 170
Baudelaire, 15
Baudrillard, Jean, 44–45
Beckett, Samuel, 16, 26, 183
Beggar's Opera, The (Gay), 80–84, 85, 87
behavioral science, 47
Bergson, Henri, 16, 17–21, 22, 58, 118
Birth of Tragedy, The (Nietzsche), 103
Bohm, David, 105
Book of the Courtier, The (Castiglione), 42
Bourdieu, Pierre, 133
Brecht, Bertold, 85–88, 102, 106, 148. *See also specific plays*
Brennan, Teresa, 27
Bright Room Called Day, A (Kushner), 172
Bruno, Giordano, 106
Butler, Judith, 6–7, 27, 91, 164–65
Byronism, 108

Caillé, Alain, 140–41
Cantor's Proof, 117.
capitalism, 133, 145, 146, 154, 163, 166, 167, 176
Carlson, Susan, 164
carnival, 3, 71, 75–76, 77, 78, 79, 82–83, 84, 88, 102, 104, 139, 152, 163, 183

cartesianism. *See* Descartes, René
Castiglione, 42
celebration, 85, 91, 95, 104
Cervantes, 72, 100
chaotics, 4, 77, 105–6, 107–12, 116, 144, 145, 153, 166, 167
Chekhov, Anton, 153
Cherry Orchard, The (Chekhov), 153
choler, 55
Churchill, Caryl, 165
citation, 6
Clouds, The (Aristophanes), 95–101
Coast of Utopia (Stoppard), 175–77, 180
cognitive sciences, 89–90, 103
Comedians (Griffiths), 170–71
commodity exchange, 133, 135, 140, 141, 145, 153, 166, 167
Congreve, William, 25
consciousness theory, 113, 114, 116, 118, 120, 155, 157–58, 179, 181
con-speciation, 45–46
Constant Wife, The (Maugham), 164
cooperation, 4, 21, 22, 120, 121–24, 153, 182, 183
Copernicus, 42
Cowell, Andrew, 132
Crane, Mary Thomas, 55
Cromwell, Oliver, 60

Dadaism, 114–15
Damasio, Antonio, 46–48, 53–54, 89–90, 101, 103, 156, 157
dark comedy, 37
Darwinism, 120, 127, 166
Dawkins, Richard, 22, 121–23, 128–29, 143, 144, 157
deconstruction, 15
Deleuze, Gilles, 5
de Man, Paul, 155, 156
Dennett, Daniel C., 90, 113
Derrida, Jacques, 6, 95, 128, 129, 131, 133, 134, 138, 147
Descartes, René, 12–13, 53, 54–55, 89–90, 101
Dionysian, 103–4, 105

dualism, mind-body, 30, 43, 44, 53, 54–55, 90; Apollonian-Dionysian, 103–4, 105
Durkheim, Émile, 44–45

Eagleton, Terry, 13
ecology, 21, 22, 183
enchantment, 44–45
environmentalism, 46–47, 113, 121, 122, 156, 158, 159, 167, 169, 171, 177, 179, 180
Euripides, 100
Every Man Out of His Humour (Jonson), 58–60, 62
evolutionarily stable systems, 120
evolutionary theory, 120, 122
exchange, economies of, 127–28, 131–34, 138, 148
Exonerated, The (Blank and Jensen), 25

Falstaff, Sir John, 62–70, 70–71, 73, 78, 91, 101, 176, 177. *See also* Shakespeare's Henry *plays*
Fielding, Henry, 72, 100
fitness, 22
Fleming, John, 106
Flieger, Jerry, 16
Frazier, Nancy, 24
Freud, Sigmund, 22, 90, 157

Galileo (Stoppard), 106–7
game theory, 122–24, 143, 144
gargantua, 112
Gay, John, 80–84, 85, 87, 88–89, 102, 148
"gay science," 13
gift, 4, 95, 128, 129–34, 136, 138, 139, 140–41, 143, 144, 147, 148, 150, 151, 167, 169, 173, 180
Gillespie, Michael, 154
Gnosticism, 12, 55
Goldsmith, Oliver, 25, 152
Gould, Stephen Jay, 21–22
Griffiths, Trevor, 170
grotesque realism, 3, 74–76, 77, 80
Gruber, William E., 8–9

Hamlet (Shakespeare), 11, 20, 40–52, 53, 60, 61, 63, 66, 90–91, 130, 179, 182
happy endings, 6, 164
Hart, Moss, 145
heat death, 111–12
Hegel, Georg, 13, 26, 27, 172
Henry IV, Parts 1 and 2 (Shakespeare), 62, 64–70
Henry V (Shakespeare), 62–63, 68, 70
Hillman, David, 55
Hobbes, Thomas, 16
Hokenson, Jan Walsh, 15–16, 18
homeodynamics, 113.
homeostasis, 91, 113
Homer, 26–27
Houseman, A. E., 124
Huizinga, Johan, 63
humanism, 5, 15, 16, 118
humor, 2–3, 170, 182
humoural physiology, 3, 9, 20, 55–59, 68, 69, 78, 103
Hyde, Lewis, 132, 138, 140

Ibsen, Henrik, 11
Iceman Cometh, The (O'Neil), 89
idealism, 3, 8, 13–14, 16, 18, 20, 21, 23, 24, 32, 33–34, 35, 38, 40, 41–42, 43, 49, 53, 59, 88, 100, 103, 106, 167, 176, 178, 179, 180, 183
Iliad, The (Homer), 26
Importance of Being Earnest, The (Wilde), 115, 154–55
Invention of Love, The (Stoppard), 127–28
irony, 155, 156

jokes, 16–17, 84, 158, 170
Jonson, Ben, 57–62, 77–80, 88, 107. *See also specific plays*
Jumpers (Stoppard), 116–20, 127–28
Juno and the Paycock (O'Casey), 1
justice, 5, 135, 136, 142, 156, 175, 176

Kaufman, George S., 145
Kellert, Stephen, 109

Kierkegaard, Søren, 95
King Lear (Shakespeare), 20, 179, 182
Kushner, Tony, 171–75

Lacan, Jacques, 17
Langer, Suzanne, 37–38, 71
laughter, 17–19, 71, 72–73, 158–59, 170
Leggatt, Alexander, 154
Life of Galileo, The (Brecht), 106
love, 124–25, 150, 151, 165
ludicrous, the, 11
ludic self, the, 63–64
Lyotard, Jean François, 5, 15
Lysistrata (Aristophanes), 8–9, 92–93, 96

Macbeth (Shakespeare), 20, 22, 179, 182
Machiavelli, Niccolo, 42, 69
Major Barbara (Shaw), 166–68
Malinowski, Bronislaw, 132
Mamet, David, 134–36
Mardi Gras, 71, 183
marketplace, the, 76, 77, 78, 81, 82–83, 94, 139, 160.
marriage, 149–52, 164, 175
Marxism, 5, 85, 88, 102, 133, 176–77
Maugham, Somerset, 164
Measure for Measure (Shakespeare), 72–73, 77, 80, 137–38
Meeker, Joseph W., 2–3, 17, 21, 23, 45
melancholia, 28, 55
memes, 22, 28, 128–29, 130, 157
Merchant of Venice, The (Shakespeare), 9, 20, 140–45, 149–50, 151, 165, 169, 174, 183
Meredith, George, 72, 74
metanarrative, 15
metatheatricality, 63
Midsummer Night's Dream, A (Shakespeare), 49, 137–38
Min, Eun Kyung, 135
Misanthrope, The (Molière), 60–62, 72
Moby Dick (Melville), 11
Molière, 25, 60–62, 72. *See also specific plays*

morphological diversity, 109–10
Much Ado About Nothing (Shakespeare), 49, 66, 95, 151–52, 165
mutability, 13
Myers, Henry, 11, 12

natural selection, 21
New Orleans, 182–83
Newtonianism, 110, 111
Nietzsche, Friedrich, 13, 15, 103–4
nonlinearity, 4
nonstereotypical responses, 46, 47
non-zero-sum measure, 4, 130, 138–39, 140, 148–52, 169
nurturance, 173, 174

O'Casey, Sean, 1
Odets, Clifford, 159–60
Odyssey, The (Homer), 26–27, 70–71
Oedipus (Sophocles), 11
Oliver, Kelly, 7, 24, 27–28, 100, 124, 156, 157, 165, 183
Olson, Kirby, 5–6
O'Neill, Eugene, 89
oppression, pathology of, 7, 8, 120, 145, 155–56, 158–59, 163, 166, 171, 173, 175, 181
Osteen, Mark, 131, 133–34, 136, 139, 144
Ostovich, Helen, 57, 59
Othello (Shakespeare), 20, 22, 179, 182

Peace (Aristophanes), 93
Peat, F. David, 105
phlegm, 55
Pinsky, L. E., 71
Plato, 12, 13, 16, 44, 46, 55, 90, 91, 93, 183
plague, the, 77, 80
Poetics, The (Aristotle). *See* Aristotle
populism, 15, 16, 76
postmodernism, 5–6, 15–16
potlatch, 131
Prince, The (Machiavelli), 42
Purdie, Susan, 16–17, 61

Rabelais, François, 3, 15, 71, 72, 74–76, 80, 84, 85, 88, 92, 93, 103, 104, 112, 130, 139
rationalism, 5, 15, 105
reciprocity, 132, 139, 143, 147
recognition, economy of, 27
Richard II (Shakespeare), 69
Richard III (Shakespeare), 22
Riddle, Dorothy (Riddle's Scales), 173–74
Rock 'n' Roll (Stoppard), 177–78
romanticism, 108, 109, 110, 139, 145
Rose, Stephen, 113
Ruling Class, The (Barnes), 168–70
rusticity, 96, 100–101

Saroyan, William, 160, 163
satire, 6, 8, 14–15, 19, 60, 85, 180
Scott, Nathan, Jr., 11–12, 26, 71
sedimentation, 6, 7
Segal, Erich, 9, 26, 40, 52, 70, 92, 94, 112
self-fabrication, 156
self-knowledge, 156
self-revelation, 156
Shakespeare, William, 11–12, 26, 71, 107. *See also specific plays*
Sharp, Ronald A., 140–41
Shaw, Bernard, 166, 167
Sheridan, Richard Brinsley, 25.
skepticism, 15
Smith, Adam, 136, 166; *The Theory of Moral Sentiment*, 134–35; *The Wealth of Nations*, 134–35
Smith, John Maynard, 120
Socrates, 12, 44, 46, 97–98, 101
speciation, 17, 21, 28, 40–41, 88, 121, 123, 153, 156, 178
Spinoza, Baruch, 89–90, 101
Stoppard, Tom, 9, 106–20, 124–25, 175, 177, 180. *See also specific plays*
Storey, Robert, 17, 21, 22
Stotts, Andrew, 182
Styan, J. L., 25–26
support, 174
Sypher, Wylie, 29–30, 70, 83–84, 103, 104

Taming of the Shrew, The (Shakespeare), 152, 165
Tempest, The (Shakespeare), 138, 139, 179
Tillyard, E. M. W., 56
Time of Your Life, The (Saroyan), 160–64
Titus Andronicus (Shakespeare), 20
tolerance, 174
Top Girls (Churchill), 165–66
Torrance, Robert W., 15, 85
tragedy, 2, 3, 11–12, 13, 15, 20, 21, 23, 24, 26, 29–30, 36–37, 38, 39, 40, 49, 52, 59, 91, 93, 95, 103
tragicomedy, 37
transformative remedies, 24, 28, 39, 154, 155, 157, 172, 180, 183
transmodernism, 16
Travesties (Stoppard), 112, 113–16
Troilus and Cressida (Shakespeare), 30–39, 40–41, 53, 61, 65, 72, 74, 165
Twain, Mark, 100

Twelfth Night (Shakespeare) 71, 73, 150–51, 152

utopia, 123, 124, 156, 175, 182

Verfremdungseffekt, 88–89
vitalism, 18, 22
Voltaire, 72

Wilbur, Richard, 60
Wilde, Oscar, 25, 124, 154–55, 158, 159, 160
Wilson, Timothy D., 156–58
Wodehouse, P. G., 154

You Can't Take It With You (Kaufman and Hart), 145–48, 167

Zeno's paradoxes, 116–17, 119, 129, 179–80
zero-sum measure, 130, 133, 137, 138–39, 140, 177

9 780230 604711